Rings and factorization

Rings and factorization

David Sharpe

Department of Pure Mathematics
University of Sheffield

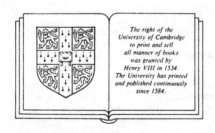

CAMBRIDGE UNIVERSITY PRESS
Cambridge
New York   New Rochelle   Melbourne   Sydney

CAMBRIDGE UNIVERSITY PRESS
Cambridge, New York, Melbourne, Madrid, Cape Town, Singapore, São Paulo, Delhi

Cambridge University Press
The Edinburgh Building, Cambridge CB2 8RU, UK

Published in the United States of America by Cambridge University Press, New York

www.cambridge.org
Information on this title: www.cambridge.org/9780521330725

© Cambridge University Press 1987

This publication is in copyright. Subject to statutory exception
and to the provisions of relevant collective licensing agreements,
no reproduction of any part may take place without the written
permission of Cambridge University Press.

First published 1987
Re-issued in this digitally printed version 2008

A catalogue record for this publication is available from the British Library

Library of Congress Cataloguing in Publication data

Sharpe, D. W. (David William)
Rings and factorization.
Bibliography
Includes index.
1. Rings (Algebra) 2. Factorization (Mathematics)
I. Title.
QA247.S525   1987   512'.4   86-33365

ISBN 978-0-521-33072-5 hardback
ISBN 978-0-521-33718-2 paperback

To my parents

CONTENTS

	Preface	ix
	Introduction	1
Part 1	**Rings**	
1.1	Introduction	5
1.2	Binary operations	5
1.3	Definition of a ring	8
1.4	Homomorphisms, subrings and isomorphisms	12
1.5	Integral domains	20
1.6	Fields	23
1.7	Residue-class rings	28
Part 2	**Factorization**	
2.1	Introduction	35
2.2	Unique factorization domains	36
2.3	Euclidean domains	39
2.4	Greatest common divisors	46
2.5	Prime elements	53
2.6	Euclidean domains are UFDs	55
2.7	The two-squares theorem	56
2.8	Factorization of polynomials	62
2.9	An application of UFDs to determinantal identities	82
	Postscript	90
	Partial solutions to exercises	92
	References	107
	Index	109

PREFACE

This little book is about rings and factorization, but it is intended to be an exhaustive study of neither of these. If such a study exists, it will have to be sought in the standard works on the subject. The present aim is more modest. The book arose out of a short series (sequence?) of 20 lectures given to second year mathematics students at the University of Sheffield. As any university teacher will testify, by no means all mathematics students feel at home with abstract ideas, let alone see the point of them. The aim of the course was to help students to make the transition into a more abstract world as painlessly as possible by presenting abstract ideas in a fairly concrete context.

I am grateful to my students for suffering the inadequacies of my exposition. It is my hope that many if not all of them gained some pleasure as well as knowledge from the course, and that readers of this book will do the same.

Exercises have been included in most sections, with an indication of how to solve most of them at the end. I appeal to readers not to cheat by looking at the answers before they have had a go themselves; this is as bad as reading a whodunnit by starting at the last page!

It is assumed that readers are familiar with the basic ideas of sets and mappings, including the notions of injective (one-one), surjective (onto) and bijective (one-one and onto) mappings, although they do not appear prominently in the book; they are principally in Section 1.4. It is also assumed that readers are familiar with the idea of an equivalence relation on a set, and of how it partitions a set into subsets. These topics will be found in a textbook on elementary set theory.

INTRODUCTION

Everybody, or nearly everybody, knows the Fundamental Theorem of Arithmetic, even if not by name. It goes back to the ancient Greek mathematicians. Start with a whole number, preferably positive, say 60. Try factorizing 60 into smaller and smaller factors until you cannot do it any further, and you end up with $60 = 2 \times 2 \times 3 \times 5$. The factors 2, 2, 3, 5 are 'irreducible' in that they cannot be further factorized. Another name for them is that they are 'prime'. Moreover, this is just about the only way of factorizing 60 into irreducibles. You could, for example, write $60 = (-2) \times 2 \times (-3) \times 5$ if you insisted, or you could change the order of the factors, but most reasonable people would say that these factorizations are no different from the previous one. Thus there is essentially one and only one way of factorizing 60 into irreducible (or prime) factors, and this is true for all whole numbers. This is the Fundamental Theorem of Arithmetic.

The same sort of thing happens with polynomials. Take the polynomial $x^3 + x^2 + x + 1$, try factorizing as far as you can, and you end up with

$$x^3 + x^2 + x + 1 = (x+1)(x^2+1).$$

At least, that is what you end up with if you insist that all the coefficients are real numbers. If you are brave enough to allow complex coefficients, you can get further, and write

$$x^3 + x^2 + x + 1 = (x+1)(x+i)(x-i).$$

Another fundamental theorem, called this time the Fundamental Theorem of Algebra, tells us that a polynomial with complex coefficients can be factorized into linear factors; and this can be done in essentially only one way. You could, for example, write

$$x^3 + x^2 + x + 1 = (ix+i)(-ix+1)(x-i),$$

but most reasonable people would say that this is the same factorization as the previous one; the first factor has been multiplied by i and the second by $-i$, which has little effect overall. This famous theorem was first proved by the great Gauss in his doctoral thesis in 1799, although the result was known before then. It is in fact not a theorem in algebra at all, but one in analysis, and Gauss's proofs (for he gave more than one) would not stand the test of today's higher standards of rigour. If the polynomial you start with has real coefficients and you insist that the factors have real coefficients, then the polynomial can only be factorized into irreducible linear and quadratic factors.

Here are two very different situations where the same thing happens, firstly with integers and secondly with polynomials. The questions naturally arise, does it happen in other situations and does it always happen? The answers are 'yes' and 'no', respectively. We shall need to set up a suitable algebraic framework in which to consider these questions, and this brings us to the idea of a 'ring'. Roughly (very roughly) speaking, to get the idea of a ring, just think what properties whole numbers (integers) and polynomials have in common, write them down and say that anything that satisfies these properties is a ring. In particular, you can add, subtract and multiply in a ring subject to some rather obvious-looking rules. The whole point of moving into this abstract setting rather than staying with the more familiar whole numbers, or polynomials, is that you can deal with both situations, and hopefully many more, at the same time.

But why bother? Isn't it just abstraction for abstraction's sake, the curse of today's unfortunate student? Hopefully not! In the first place, a number of results about whole numbers come out of this abstraction in a very elegant way. But, more importantly, factorization in this more abstract setting was forced upon mathematicians by perhaps the most famous unsolved problem of all in mathematics, Fermat's Last 'Theorem' (in inverted commas because no one has managed to prove or disprove it, yet). We all know that $3^2 + 4^2 = 5^2$, so that there are positive integers x, y, z such that $x^2 + y^2 = z^2$. The French mathematician Pierre Fermat proposed in about 1637 that, when $n > 2$, there are no positive integers x, y, z such that

Introduction 3

$x^n + y^n = z^n$. He wrote in the margin of his copy of Bachet's translation of Diophantus' *Arithmetica* that he had discovered the most remarkable short proof of this result, but that the margin was too small to contain it. Ever since, mathematicians, professional and amateur alike, have tried and failed to prove it. Gauss gave a proof for $n = 3$ which involves the idea of factorization in a ring other than the two mentioned so far. In 1843, Kummer gave what he thought was a proof of the general result, but Dirichlet pointed out to him that he had assumed that unique factorization held in a particular setting in which it did not in fact hold. It was this that brought out the importance of factorization in general, and led Dedekind and others to restore unique factorization not for single elements any more but for whole sets of elements called 'ideals'. But this takes us beyond the scope of this little book. We can only hope that this book will whet readers' appetites for further study.

Factorization of integers into their prime factors has recently come into prominence in a surprising way. Security-conscious governments are naturally concerned to be able to pass messages without their being intercepted, and this is done in code, hopefully in a code which cannot be broken by an 'enemy'. Such a coding technique is provided by the so-called 'Public-Key Cryptography'. The receiver of the message starts with two (or more) large prime numbers, say P, Q, having of the order of 50 to 100 digits. He gives the sender of the message only the product PQ, which the sender uses to encode the given message. The message can only be decoded when the individual primes P, Q are known. Herein lies the difficulty in breaking the code. The time required to factorize a large number into its prime factors increases exponentially with the size of the number, unlike that required to test that a given integer is prime, which only increases linearly with the size of the number. Thus, even in these days of very large computers, it is not practicable to extract the individual factors P, Q just from a knowledge of their product PQ. Thus, at the moment, and unless you are very unlucky (i.e. your 'enemy' is very lucky), your code is secure. This puts factorization of numbers in the forefront of research by today's computer scientists and makes it of interest to more than undergraduates in mathematics! For a popular article on this subject, reprinted from *The*

Guardian, we refer readers to 'Prime Numbers and Secret Codes' by Keith Devlin, in *Mathematical Spectrum*, where references to research articles may be found[†].

But now it is time to come down from these dizzy heights and get down to details.

[†] Added in proof. It has just been announced that H. W. Lenstra, University of Amsterdam, has devised a technique which makes the factorization of large numbers more possible, so maybe security services the world over will have to think again!

Part 1

Rings

1.1

Introduction

The first mathematics that most of us met concerned whole numbers or integers,

$$\ldots, -3, -2, -1, 0, 1, 2, 3, \ldots.$$

We learned how to add, subtract and multiply them. Later on, we thought how to divide them, but that introduces fractions so we will lay it aside for the moment. With increased sophistication, we dealt with polynomials such as $x^2 + 2$, $x^3 - 3x^2 + 1$, and added, subtracted and multiplied these as well. Both of these systems satisfy the same simple laws for addition, subtraction and multiplication. In fact, these laws are usually used without thinking. However, if we are to consider other systems than these, we shall need to make a note of the properties that we need and, in theory at any rate, we should check that the properties are satisfied each time a new system is introduced. These properties are incorporated in the definition of a ring, first formally given by A. A. Fraenkel in 1914.

1.2

Binary operations

First the idea of a binary operation on a set. Let S be a non-empty set. A *binary operation* on S is a rule whereby, given elements a, b of S (which could be the same element), there is defined a unique element of S, denoted variously by $a + b$ or ab or $a \cdot b$ or $a \circ b$ according to how the binary operation is denoted. For example, if $\mathbb{Z}$ denotes the set of integers (as it always does following the German word '*zahlen*', meaning 'number'), then addition and multiplication are both binary operations on $\mathbb{Z}$ in that, given integers a, b, there are defined unique integers $a + b, ab$ in $\mathbb{Z}$. If we are considering the binary operation $+$ on $\mathbb{Z}$, we sometimes but not always write $(\mathbb{Z}, +)$ to emphasize this, and $(\mathbb{Z}, \cdot)$ or $(\mathbb{Z}, \times)$ when we are considering multiplication on $\mathbb{Z}$. If we are considering both addition and multiplication on $\mathbb{Z}$, we may write $(\mathbb{Z}, +, \cdot)$. Generally, if we have a non-empty set S and a binary operation denoted by $+$ defined on S, we write $(S, +)$; and $(S, +, \cdot)$ will denote S with two binary operations on it, denoted by addition and multiplication. We emphasize that the elements of S may not be numbers and that the binary operations, although they may be denoted by addition and multiplication, may have nothing to do with addition and multiplication of numbers. Surprisingly, it is less confusing to stick to additive and multiplicative notation rather than to introduce such devices as ∘ and ∗ to denote a binary operation on a set.

This definition of a binary operation on S may leave you feeling a little dissatisfied or cheated. After all, what is a 'rule'? For the more sophisticated reader, a binary operation on S is a mapping from the Cartesian product $S \times S$ to S. The Cartesian product $S \times S$ consists of all ordered pairs (a, b), where a, b are elements of S (similar to the Cartesian coordinates of points in a plane), and the image of the pair (a, b) under the mapping is denoted by $a + b$ or ab or $a \circ b$ according to the way in which the binary operation is designated. But the sophisticated way of thinking of a binary operation is seldom the best, so we shall resort to the simple-minded way.

The examples which follow are all examples of well-known sets with two binary operations defined on them. As well as illustrating the notion of a binary operation, they will also serve to establish some standard notation for various sets.

Examples of sets with binary operations
   (1) $(\mathbb{Z}, +, \cdot)$, where $\mathbb{Z}$ is the set of integers.
   (2) $(\mathbb{Q}, +, \cdot)$, where $\mathbb{Q}$ is the set of rational numbers.
   (3) $(\mathbb{R}, +, \cdot)$, where $\mathbb{R}$ is the set of real numbers.
   (4) $(\mathbb{C}, +, \cdot)$, where $\mathbb{C}$ is the set of complex numbers.
   (5) $(M_n(\mathbb{R}), +, \cdot)$, where $M_n(\mathbb{R})$ denotes the set of all $n \times n$ matrices with real entries, and $+, \cdot$ denote the usual matrix addition and multiplication.

1.2 Binary operations

(6) $(\mathbb{R}[x], +, \cdot)$, where $\mathbb{R}[x]$ denotes the set of all polynomials in x with real coefficients and $+, \cdot$ denote the usual addition and multiplication of polynomials.

(7) $(\mathbb{R}[[x]], +, \cdot)$, where $\mathbb{R}[[x]]$ denotes the set of all 'formal power series'. These are formal expressions of the form

$$a_0 + a_1 x + a_2 x^2 + \ldots,$$

where the coefficients $a_0, a_1, a_2, \ldots$ are all real numbers. Although this looks like an infinite sum, it is not, and no concept of convergence is involved, as would be the case if we were trying to add infinitely many things together. The formal expression really stands for the infinite sequence $(a_0, a_1, a_2, \ldots)$, and the powers of x are merely used to denote the positions in the sequence. It should be noted that

$$\sum_{n=0}^{\infty} a_n x^n = \sum_{n=0}^{\infty} b_n x^n,$$

where the a_n and b_n are real numbers, if and only if $a_n = b_n$ for every n. Addition and multiplication on $\mathbb{R}[[x]]$ are defined by

$$\left(\sum_{n=0}^{\infty} a_n x^n\right) + \left(\sum_{n=0}^{\infty} b_n x^n\right) = \sum_{n=0}^{\infty} (a_n + b_n) x^n,$$

$$\left(\sum_{n=0}^{\infty} a_n x^n\right)\left(\sum_{n=0}^{\infty} b_n x^n\right) = \sum_{n=0}^{\infty} \left(a_0 b_n + a_1 b_{n-1} + \ldots + a_n b_0\right) x^n.$$

These rules are similar to the rules for the addition and multiplication of polynomials. In fact, $\mathbb{R}[x]$ is a subset of $\mathbb{R}[[x]]$ in that a polynomial is just a power series with only finitely many non-zero coefficients.

(8) Instead of considering polynomials in a single variable x, we can have polynomials in n independent variables (usually called 'indeterminates') $x_1, \ldots, x_n$ with real coefficients (say). We denote the set of these by $\mathbb{R}[x_1, \ldots, x_n]$. Such a polynomial is a formal expression of the form

$$\sum a_{i_1 \ldots i_n} x_1^{i_1} x_2^{i_2} \ldots x_n^{i_n},$$

where $i_1, \ldots, i_n$ are non-negative integers and $a_{i_1 \ldots i_n} \in \mathbb{R}$[†] with only

---

[†] The symbol $\in$ stands for 'belongs to'. Thus $a \in S$ means that a is an element of the set S. Also, $a \notin S$ means that a is not an element of S.

8   Rings

finitely many of the $a_{i_1\ldots i_n}$ non-zero. (For instance, an example of an element of $\mathbb{R}[x_1, x_2, x_3]$ is $4x_1^2x_2^3x_3 - 6x_2x_3^4 + x_1^4$; this is also an element of the set $\mathbb{Z}[x_1, x_2, x_3]$ since its coefficients are actually integers.) Addition and multiplication can be defined on $\mathbb{R}[x_1, \ldots, x_n]$ by

$$(\sum a_{i_1\ldots i_n} x_1^{i_1}\ldots x_n^{i_n}) + (\sum b_{i_1\ldots i_n} x_1^{i_1}\ldots x_n^{i_n})$$
$$= \sum (a_{i_1\ldots i_n} + b_{i_1\ldots i_n}) x_1^{i_1}\ldots x_n^{i_n},$$
$$(\sum a_{i_1\ldots i_n} x_1^{i_1}\ldots x_n^{i_n})(\sum b_{i_1\ldots i_n} x_1^{i_1}\ldots x_n^{i_n})$$
$$= \sum (\sum_{\substack{j_r + k_r = i_r \\ 1 \leq r \leq n}} a_{j_1\ldots j_n} b_{k_1\ldots k_n}) x_1^{i_1}\ldots x_n^{i_n}.$$

Examples of these rather complicated looking rules would be

$$(4x_1^2x_2^3x_3 - 3x_2x_3^4 + x_1^4) + (x_1^2x_2^3x_3 + 3x_2x_3)$$
$$= 5x_1^2x_2^3x_3 - 3x_2x_3^4 + x_1^4 + 3x_2x_3,$$
$$(4x_1^2x_2^3x_3 - 3x_2x_3 + x_1^4)(x_1^2x_2^3x_3 + 3x_2x_3)$$
$$= 4x_1^4x_2^6x_3^2 + (12-3)x_1^2x_2^4x_3^2 - 9x_2^2x_3^2$$
$$+ x_1^6x_2^3x_3 + 3x_1^4x_2x_3.$$

(9) $(C[a, b], +, \cdot)$, where $C[a, b]$ denotes the set of all continuous real-valued functions defined on a closed interval $[a, b]$ and $+, \cdot$ are defined pointwise, i.e.

$$(f+g)(x) = f(x) + g(x), (fg)(x) = f(x)g(x).$$

Exercise 1.2 Think of ten other examples of sets with one or two binary operations defined on them. Try to make them as different as you can.

1.3

Definition of a ring

As we list the axioms that must be satisfied by a ring, it is suggested that readers bear in mind the prototype examples of the integers and polynomials. In fact, all the examples listed in Section 1.2 of sets with two binary operations will satisfy the axioms and so are rings.

1.3 Definition of a ring

Definition 1.3.1 A 'ring' is a non-empty set R which satisfies the following axioms:
(1) R has a binary operation denoted by $+$ defined on it;
(2) addition is associative, i.e.

$$a+(b+c)=(a+b)+c \text{ for all } a, b, c \in R$$

(so that we can write $a+b+c$ without brackets);
(3) addition is commutative, i.e.

$$a+b=b+a \text{ for all } a, b \in R;$$

(4) there is an element denoted by 0 in R such that

$$0+a=a \text{ for all } a \in R$$

(there is only one such element because, if $0_1, 0_2$ are two such, then $0_1 = 0_1 + 0_2 = 0_2$ and they are the same – we call 0 the *zero element* of R);
(5) for every $a \in R$, there exists an element $-a \in R$ such that

$$(-a)+a=0$$

(there is only one such element for each a, because if $b+a=0$ and $c+a=0$, then

$$b=0+b=(c+a)+b=c+(a+b)=c+0=c;$$

we call $-a$ the *negative* of a);
(6) R has a binary operation denoted by multiplication defined on it;
(7) multiplication is associative, i.e.

$$a(bc)=(ab)c \text{ for all } a, b, c \in R;$$

(8) multiplication is left and right distributive over addition, i.e.

$$a(b+c)=ab+ac, (a+b)c=ac+bc \text{ for all } a, b, c \in R;$$

(9) there is an element denoted by 1 in R such that $1 \neq 0$ and

$$1 \cdot a = a \cdot 1 = a \text{ for all } a \in R$$

(as for the zero element, there is only one such element, and it is called the *identity element* of R).

Axioms 1–5 may be summarized by saying that R is an abelian group under addition.

Axiom 9 is not imposed by all authors. Thus, for example, the even integers with the usual addition and multiplication would form a ring without an identity element. However, we shall insist that our rings possess identity elements and that $1 \neq 0$. It may be asked how 1 and 0 could be equal. If we take a single-element set $\{x\}$ and define addition and multiplication on it by $x + x = x$, $xx = x$, then we obtain a 'ring' in which $0 = x = 1$. Such a ring, with only one element, is called a *trivial ring*. In fact, to anticipate Theorem 1.3.2, if $1 = 0$ in a ring, then $a = a \cdot 1 = a \cdot 0 = 0$ for all a, and such a ring must be a trivial ring. Thus trivial rings provide the only situation in which $1 = 0$. Thus trivial rings are excluded by our Axiom 9.

The next result includes some elementary consequences of the definition which are usually used without thinking.

Theorem 1.3.2 Let R be a ring. Then

(1) $0a = a0 = 0$ for all $a \in R$,
(2) $(-a)b = a(-b) = -(ab)$ for all $a, b \in R$,
(3) $(-a)(-b) = ab$ for all $a, b \in R$.

Proof (1) $0a + 0a = (0 + 0)a = 0a$[†], so that

$$-(0a) + (0a + 0a) = -(0a) + 0a,$$
$$(-(0a) + 0a) + 0a = 0,$$
$$0 + 0a = 0,$$
$$0a = 0.$$

A similar argument shows that $a0 = 0$.

(2) $(-a)b + ab = ((-a) + a)b = 0b = 0$ by (1), so that $(-a)b = -(ab)$. A similar argument shows that $a(-b) = -(ab)$.

(3) $(-a)(-b) = -(a(-b))$ by (2)
$\qquad\qquad = -(-(ab))$ by (2)
$\qquad\qquad = ab$ by Axiom 5. $\square$

---

† The usual conventions concerning addition and multiplication apply. Thus $0a + 0a$ means $(0a) + (0a)$.

1.3 Definition of a ring

All the rings listed in Section 1.2 except the matrix rings have an extra property incorporated in the following definition:

Definition 1.3.3 A ring R whose multiplication is commutative, i.e. such that $ab = ba$ for all $a, b \in R$, is said to be a 'commutative ring'.

Of course, for a commutative ring the left and right distributive laws amount to the same thing.

Although it is possible to consider factorization in non-commutative rings, it is very much harder, and is well beyond the scope of this book. Thus we shall be concerned almost exclusively with commutative rings.

Now that we have defined an arbitrary commutative ring, we can immediately see how the construction of polynomials in one or more variables, or of power series, as given in Section 1.2, can be carried out when the coefficients belong to an arbitrary commutative ring R. We can define a power series in a 'variable' x (more correctly referred to as an *indeterminate* x), with coefficients in R, to be a formal expression of the form

$$a_0 + a_1 x + a_2 x^2 + \ldots,$$

where $a_0, a_1, a_2, \ldots$ belong to R. The addition and multiplication of power series are defined as in Section 1.2, and these make the power series into a commutative ring, denoted by $R[[x]]$. Its zero element is

$$0 + 0x + 0x^2 + \ldots$$

and its identity element is

$$1 + 0x + 0x^2 + \ldots,$$

where 0, 1 denote the zero and identity elements of R, respectively. It is usual also to denote the two power series by 0, 1, respectively. If we consider only those power series with only finitely many non-zero coefficients, and use the same addition and multiplication, we obtain the *polynomial ring* $R[x]$, again a commutative ring. And a similar thing can be done with more than one variable (or indeterminate), to obtain the polynomial ring in n variables over R, written $R[x_1, \ldots, x_n]$.

12   Rings

The formation of polynomial rings is a very useful 'new rings from old' construction. Another such construction is to form $n \times n$ matrices with entries in a commutative ring R. This will give the matrix ring $M_n(R)$. We shall not use this construction here, however, because it will produce a non-commutative ring (except when $n = 1$).

Exercises 1.3

1. Let R, S be rings. Show that the cartesian product $R \times S$ consisting of all ordered pairs (a, b), where $a \in R$, $b \in S$, is a ring with addition and multiplication defined by

   $(a, b) + (c, d) = (a + c, b + d)$,

   $(a, b)(c, d) = (ac, bd)$

   for all $a, c \in R, b, d \in S$.

2. Let R be a ring. Show that the cartesian product $R \times R$ is a ring with addition and multiplication defined by

   $(a, b) + (c, d) = (a + c, b + d)$,

   $(a, b)(c, d) = (ac, ad + bc)$

   for all $a, b, c, d \in R$.

3. The ring R has the property that $a^2 = a$ for every element a of R. (Such a ring is called a *Boolean ring* after George Boole.) Show that $a = -a$ for every $a \in R$ and that R is commutative.

4. For a non-empty set S, denote by $P(S)$ the power set of S, i.e. the set of all subsets of S. For $X, Y \in P(S)$, define the *symmetric difference* $X \triangle Y$ by

   $X \triangle Y = (X \setminus Y) \cup (Y \setminus X)$.

   ($X \setminus Y = \{x \in X : x \notin Y\}$). Show that $(P(S), \triangle, \cap)$ is a ring which is actually a Boolean ring.

1.4

Homomorphisms, subrings and isomorphisms

Let R be a commutative ring. A typical member of the polynomial ring $R[x]$ is a formal expression of the form

$$f(x) = a_0 + a_1 x + a_2 x^2 + \cdots$$

1.4 Homomorphisms, subrings and isomorphisms

with only finitely many non-zero coefficients. Let α be an element of R. We can 'evaluate' this polynomial at α to give the element

$$f(\alpha) = a_0 + a_1\alpha + a_2\alpha^2 + \ldots$$

of R. Note that, although this sum appears infinite, it is not because only finitely many of the a_i's are non-zero. Thus $f(\alpha)$ is a perfectly good element of R. Although $f(x)$ and $f(\alpha)$ are very different things (after all, they belong to different rings, $f(x)$ to $R[x]$ and $f(\alpha)$ to R), yet their addition and multiplication are similar:

$$\left.\begin{array}{l}\left(\sum_{n=0}^{\infty} a_n x^n\right) + \left(\sum_{n=0}^{\infty} b_n x^n\right) = \sum_{n=0}^{\infty} (a_n + b_n)x^n, \\ \left(\sum_{n=0}^{\infty} a_n x^n\right)\left(\sum_{n=0}^{\infty} b_n x^n\right) = \sum_{n=0}^{\infty} (a_0 b_n + a_1 b_{n-1} + \ldots + a_n b_0)x^n\end{array}\right\} \text{in } R[x],$$

$$\left.\begin{array}{l}\left(\sum_{n=0}^{\infty} a_n \alpha^n\right) + \left(\sum_{n=0}^{\infty} b_n \alpha^n\right) = \sum_{n=0}^{\infty} (a_n + b_n)\alpha^n, \\ \left(\sum_{n=0}^{\infty} a_n \alpha^n\right)\left(\sum_{n=0}^{\infty} b_n \alpha^n\right) = \sum_{n=0}^{\infty} (a_0 b_n + a_1 b_{n-1} + \ldots + a_n b_0)\alpha^n\end{array}\right\} \text{in } R.$$

The addition and multiplication in $R[x]$ is a matter of definition, whereas that in R follows by virtue of the associative, commutative and distributive laws in R. What we have is a mapping

$$R[x] \to R$$

defined by $f(x) \mapsto f(\alpha)$, which we call *evaluation at* α. Further, this mapping has the properties incorporated in the following definition:

Definition 1.4.1 Let $f: S \to R$ be a mapping from the ring S to the ring R. We say that f is a 'homomorphism' if
(1) $f(a+b) = f(a) + f(b)$ for all $a, b \in S$,
(2) $f(ab) = f(a)f(b)$ for all $a, b \in S$,
(3) $f(1_S) = 1_R$.

We can extend the idea of evaluation to polynomials in many variables. Thus, for example, consider $f(x_1, \ldots, x_n)$ in the polynomial ring $R[x_1, \ldots, x_n]$ in n independent variables over the commutative ring R, and let $\alpha_1, \ldots, \alpha_n \in R$. If

$$f(x_1, \ldots, x_n) = \sum a_{i_1 \ldots i_n} x_1^{i_1} \ldots x_n^{i_n},$$

then we write
$$f(\alpha_1,\ldots,\alpha_n)=\sum a_{i_1\ldots i_n}\alpha_1^{i_1}\ldots\alpha_n^{i_n},$$
and we can define a homomorphism
$$R[x_1,\ldots,x_n]\to R$$
by $f(x_1,\ldots,x_n)\mapsto f(\alpha_1,\ldots,\alpha_n)$. This is referred to as 'evaluation at $(\alpha_1,\ldots,\alpha_n)$'.

In condition (3) of Definition 1.4.1, 1_S, 1_R denote the identity elements of S and R, respectively. If f is surjective, then (3) is redundant. For then, given $r\in R$, there exists $s\in S$ such that $f(s)=r$, and
$$f(1_S)r=f(1_S)f(s)=f(1_Ss)=f(s)=r.$$
By a similar argument, $rf(1_S)=r$ for all $r\in R$. Thus $f(1_S)=1_R$. But, in general, (3) does not follow from (1), (2). For example, let R be the ring $M_2(\mathbb{R})$ of all 2×2 matrices with real entries, and let S be the ring of all 2×2 matrices of the form
$$\begin{bmatrix} a & 0 \\ 0 & 0 \end{bmatrix},$$
where $a\in\mathbb{R}$, its addition and multiplication being matrix addition and multiplication. Let $f:S\to R$ be the inclusion mapping, so that $f(s)=s$ for every $s\in S$. Then conditions (1), (2) hold, but the identity element of S, namely
$$\begin{bmatrix} 1 & 0 \\ 0 & 0 \end{bmatrix},$$
does not map to the identity element of R, which is
$$\begin{bmatrix} 1 & 0 \\ 0 & 1 \end{bmatrix}.$$
This mapping is not a homomorphism according to our definition.

On the other hand, if we take the same ring R and take as S the ring $M_2(\mathbb{Q})$ of all 2×2 matrices with rational entries, then now the inclusion mapping $S\to R$ *is* a homomorphism; conditions (1), (2) of the definition hold, and the identity element of S is the same as the identity element of R, and so maps to it under the inclusion mapping.

1.4 Homomorphisms, subrings and isomorphisms

Other examples where the inclusion mapping is a homomorphism of rings are $\mathbb{Q} \to \mathbb{R}$, $\mathbb{R} \to \mathbb{C}$, $R[x] \to R[[x]]$ (where R is any commutative ring), with the usual addition and multiplication. This situation, where one ring lies inside another, is a common occurrence, so we make the following definition:

Definition 1.4.2 Let R, S be rings such that $S \subseteq R$ (i.e. S is a subset of R). If the inclusion mapping $S \to R$ is a homomorphism, then we say that S is a 'subring' of R.

We may paraphrase this definition by saying that a ring S is a subring of a ring R if $S \subseteq R$ and

$$\left. \begin{array}{c} a +_S b = a +_R b \\ a \cdot_S b = a \cdot_R b \end{array} \right\} \text{for all } a, b \in S$$
$$1_S = 1_R$$

i.e., if the addition and multiplication of S are the same as the addition and multiplication of R (on the elements of S), and if S, R have the same identity elements.

This form of the definition of a subring is not very convenient to handle, so we shall effectively recast it after we have proved the next result. This tells us how the zero element and negatives behave under homomorphisms.

Theorem 1.4.3 Let $f: S \to R$ be a homomorphism of rings. Then
 (1) $f(0_S) = 0_R$,
where 0_S, 0_R are the zero elements of S, R, respectively,
 (2) $f(-a) = -f(a)$ for all $a \in S$, where $-a, -f(a)$ denote the negatives of $a, f(a)$ in S, R, respectively.

If we take f to be an inclusion mapping in this theorem, it tells us that, if S is a subring of R, then S, R have the same zero element, and the negative of an element of S is the same whether it is regarded as an element of S or of R.

Proof of Theorem 1.4.3
 (1) $f(0_S) + f(0_S) = f(0_S + 0_S) = f(0_S)$.

If we add $-f(0_S)$ to both sides, we obtain

$$0_R + f(0_S) = 0_R,$$

which gives

$$f(0_S) = 0_R.$$

(2) Let $a \in S$. Then

$$f(a) + f(-a) = f(a + (-a)) = f(0_S) = 0_R,$$

from which $f(-a) = -f(a)$. □

We now consider the question of whether a given subset of a given ring is a subring. A convenient test of this is contained in the following result. Note the absence of any reference to the associative, commutative and distributive laws in the conditions.

Theorem 1.4.4 (The Subring Criterion) Let R be a ring and let S be a subset of R. Then S is a subring of R if and only if the following conditions hold:

(1) $1 \in S$;
(2) whenever $a, b \in S$, then $a + b \in S$;
(3) whenever $a \in S$, then $-a \in S$;
(4) whenever $a, b \in S$, then $ab \in S$.

In condition (1), 1 of course denotes the identity element of R. We may summarize the conditions by saying that S contains the identity element of R and is closed under addition, negation and multiplication.

Proof of the Subring Criterion First suppose that S is a subring of R. Then conditions (1) (2), (4) hold by definition and condition (3) holds because of Theorem 1.4.3.

Conversely, suppose that conditions (1)–(4) hold. Conditions (2), (4) enable us to use the addition and multiplication of R as binary operations on S. They are associative on S, the addition is commutative on S, and the left and right distributive laws hold in S, because they do in R. By condition (1), the identity element 1 of R belongs to S, and it clearly also acts as identity element of S. Now $1 \in S$ and so

1.4 Homomorphisms, subrings and isomorphisms

also $-1 \in S$ (by (3)), so $0 = 1 + (-1) \in S$ (by (2)). Thus the zero element of R also belongs to S, and clearly acts as zero element of S. Note also that $1 \neq 0$. Finally, given $a \in S$, then $-a \in S$, and this clearly acts as the negative of a in S as well as in R. We have now checked that all the axioms are satisfied for S to be a ring using the addition and multiplication of R, and R, S have the same identity element. Thus S is a subring of R. □

We can illustrate the use of the Subring Criterion in the following result:

Theorem 1.4.5 Let $f: S \to R$ be a homomorphism of rings. Then the image, $\operatorname{Im} f$, is a subring of R. (Note that $\operatorname{Im} f = \{f(a): a \in S\}$.)

Proof Firstly, $1_R = f(1_S) \in \operatorname{Im} f$. Now consider $a, b \in \operatorname{Im} f$. Then $a = f(s_1), b = f(s_2)$ for some $s_1, s_2 \in S$, so

$$a + b = f(s_1) + f(s_2) = f(s_1 + s_2) \in \operatorname{Im} f,$$
$$-a = -f(s_1) = f(-s_1) \in \operatorname{Im} f,$$
$$ab = f(s_1)f(s_2) = f(s_1 s_2) \in \operatorname{Im} f.$$

Hence, by the Subring Criterion, $\operatorname{Im} f$ is a subring of R. □

As an illustration of Theorem 1.4.5, we can use the idea of evaluating polynomials, although in a slightly modified form. Consider, for example, the ring $\mathbb{Z}[x]$ of polynomials in x with integer coefficients. We can define a mapping

$$\mathbb{Z}[x] \to \mathbb{R}$$

by evaluating each polynomial $f(x)$ at $\sqrt{5}$ (say). Thus $f(x)$ maps to $f(\sqrt{5})$. This is a homomorphism. We denote its image by $\mathbb{Z}[\sqrt{5}]$, so that $\mathbb{Z}[\sqrt{5}]$ is a subring of $\mathbb{R}$, and so is a ring using the addition and multiplication of real numbers. A typical element of $\mathbb{Z}[\sqrt{5}]$ is

$$a_0 + a_1\sqrt{5} + a_2(\sqrt{5})^2 + \ldots,$$

where $a_0, a_1, a_2, \ldots \in \mathbb{Z}$ with only finitely many of them non-zero, and this simplifies to a real number of the form $a + b\sqrt{5}$, where $a, b \in \mathbb{Z}$. Indeed, we can form the rings $\mathbb{Z}[\sqrt{n}]$ for all positive integers

n in a similar way. Another example would be to take the mapping

$$\mathbb{Z}[x] \to \mathbb{C}$$

defined by evaluating each polynomial at the complex number i. The image of this homomorphism is a subring of $\mathbb{C}$, denoted by $\mathbb{Z}[i]$, and consists of all complex numbers of the form $a+ib$, where $a, b \in \mathbb{Z}$. These are the so-called *Gaussian integers*. In a similar way we can form the rings $\mathbb{Z}[\sqrt{-n}]$ for all positive integers n.

Definition 1.4.6 A homomorphism which is also bijective is called an 'isomorphism'. If R, S are rings and there exists an isomorphism $f: S \to R$, then we say that the rings R, S are 'isomorphic', and write $R \approx S$.

The word 'isomorphism' comes from the Greek words *isos* = like and *morphe* = shape. For a ring R, the identity mapping $R \to R$ is an isomorphism, so $R \approx R$. Also, if $f: S \to R$ is an isomorphism, then so is the inverse mapping $f^{-1}: R \to S$. Further, if $f: S \to R$, $g: T \to S$ are isomorphisms of rings, then it is not a difficult matter to verify that the composed mapping $f \circ g: T \to R$ (given by $(f \circ g)(a) = f(g(a))$ for $a \in T$) is also an isomorphism, so that, if $R \approx S$ and $S \approx T$, then $R \approx T$. Thus, isomorphism on the class of rings is an equivalence relation. Two rings which are isomorphic have identical ring-theoretic properties.

Definition 1.4.7 A homomorphism which is also injective is called an 'embedding'. If R, S are rings and there is an embedding $f: S \to R$, then we say that the ring S is 'embedded' in the ring R.

If the ring S is embedded in the ring R, with $f: S \to R$ as an embedding, then Imf is a subring of R and $S \approx$ Imf. Sometimes one may loosely say that S *is* a subring of R. For example, given a commutative ring R, we can embed R in the polynomial ring $R[x]$ by mapping $a \in R$ to the polynomial $a + 0x + 0x^2 + \ldots$ (sometimes called, rather inaccurately, a 'constant polynomial'). It is usual to identify a and $a + 0x + 0x^2 + \ldots$ and say that R *is* a subring of $R[x]$.

1.4 Homomorphisms, subrings and isomorphisms

There are various ways of avoiding some tedious verifications when we need to show that certain structures are rings. For example, suppose we have a ring R and a *set* S such that there is a bijection $f: S \to R$. We can use this bijection to transfer the ring structure of R to S. We define addition and multiplication on S by

$$a+b = f^{-1}(f(a)+f(b)),$$
$$ab = f^{-1}(f(a)f(b)),$$

where $a, b \in S$ and where the addition and multiplication on the right-hand side are in the ring R. It is now possible to verify the ring axioms for S. In fact, the mapping f will be an isomorphism from S to R. Examples of the use of this device may be found in Exercise 1.4.5 and in the exercises in Section 1.6.

Exercises 1.4.

1. Let p be a prime number in $\mathbb{Z}$. Show that the set of all rational numbers of the form m/n, where $m, n \in \mathbb{Z}$ and p does not divide n, is a ring under the addition and multiplication of rational numbers.
2. Show that the matrices of the form
$$\begin{bmatrix} a & 0 \\ b & c \end{bmatrix},$$
where a, b, c are integers, form a ring under matrix addition and multiplication.
3. Show that the complex numbers of the form $a + b\omega$, where $a, b \in \mathbb{Z}$ and $\omega = e^{2\pi i/3}$, form a ring under the addition and multiplication of complex numbers.
4. Let $n \equiv 1 \pmod{4}$ be a fixed integer. Show that the set of all complex numbers of the form $a + \tfrac{1}{2}b(\sqrt{(n)} - 1)$, where $a, b \in \mathbb{Z}$, is a ring under the addition and multiplication of complex numbers.
5. Let $(R, +, \cdot)$ be a ring. Define binary operations $\oplus, \odot$ on R such that $(R, \oplus, \odot)$ is a ring isomorphic to $(R, +, \cdot)$ and in which the zero element and the identity element are reversed.

1.5

Integral domains

In many rings it is perfectly possible to have two non-zero elements whose product is zero. For example, in $M_2(\mathbb{R})$,

$$\begin{bmatrix} 0 & 1 \\ 0 & 0 \end{bmatrix} \begin{bmatrix} 1 & 0 \\ 0 & 0 \end{bmatrix} = \begin{bmatrix} 0 & 0 \\ 0 & 0 \end{bmatrix}.$$

Again, in the ring $C[0, 1]$ of continuous real-valued functions on $[0, 1]$, the functions f, g illustrated in the figure are non-zero, but their (pointwise) product is zero. Such a thing cannot happen in the ring $\mathbb{Z}$, however, nor in $\mathbb{Q}, \mathbb{R}, \mathbb{C}$. When the ring in question possesses elements $a, b \neq 0$ such that $ab = 0$, we say that it possesses *proper divisors of zero*. The notion of a divisor will be developed further in Part 2.

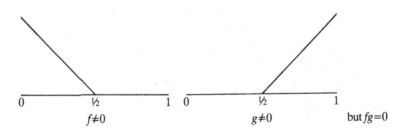

The easiest development of a theory of factorization, which is our aim, is in a ring which, like our prototype example $\mathbb{Z}$, has no proper divisors of zero, and which, like $\mathbb{Z}$, is commutative.

We therefore make the following definition:

Definition 1.5.1 A ring R is said to be an 'integral domain' if (1) it is commutative, and (2) it has no proper divisors of zero, so that

$$ab = 0 \Rightarrow a = 0 \text{ or } b = 0 \, (a, b \in R).$$

An alternative description of an integral domain is that it is a ring R which (1)' is commutative, and (2)' satisfies the cancellation law, i.e.

$$ab = ac \text{ and } a \neq 0 \Rightarrow b = c \, (a, b, c \in R).$$

1.5 Integral domains

To see that these two descriptions of an integral domain are equivalent, first let R be a commutative ring satisfying condition (2), and assume that $ab = ac$ with $a \neq 0$. Then $a(b-c) = 0$ and $a \neq 0$, so $b - c = 0$ by (2), i.e. $b = c$†. Next let R be a commutative ring satisfying condition (2)′, and assume that $ab = 0$ with $a \neq 0$. Then $ab = a \cdot 0$ so, by the cancellation law, $b = 0$. Thus (2) holds.

Besides $\mathbb{Z}$, the other prototype examples of rings where unique factorization occurs are the polynomial rings $\mathbb{C}[x]$, $\mathbb{R}[x]$. We therefore need to know that such rings are integral domains. We shall establish the following general result.

Theorem 1.5.2 Let R be an integral domain. Then the polynomial ring $R[x]$ is also an integral domain.

Proof Firstly, R is commutative, and it is easy to see that $R[x]$ is also commutative. Now consider $f(x), g(x) \in R[x]$, both non-zero. We write

$$f(x) = a_0 + a_1 x + a_2 x^2 + \ldots,$$
$$g(x) = b_0 + b_1 x + b_2 x^2 + \ldots,$$

where the a_i, b_j belong to R. There exist non-negative integers m, n such that $a_m \neq 0$, $a_{m+1} = a_{m+2} = \ldots = 0$ and $b_n \neq 0$, $b_{n+1} = b_{n+2} = \ldots = 0$. Then the coefficient of x^{m+n} in $f(x)g(x)$ is $a_m b_n$, which is non-zero, and the coefficients of higher powers of x are all zero. Thus $f(x)g(x) \neq 0$. This shows that $R[x]$ has no proper divisors of zero. □

If we define the *degree* of a non-zero polynomial $a_0 + a_1 x + a_2 x^2 + \ldots$ to be the largest integer m such that $a_m \neq 0$, then the proof of Theorem 1.5.2 shows that, when R is an integral domain,

$$\deg(f(x)g(x)) = \deg f(x) + \deg g(x)$$

for non-zero polynomials $f(x), g(x)$ in $R[x]$. This formula can be extended to apply when $f(x)$ or $g(x)$ is the zero polynomial if we give the zero polynomial the formal degree $-\infty$ and adopt the

---

† $b - c$ stands for $b + (-c)$.

conventions

$$n + (-\infty) = (-\infty) + n = -\infty$$

for all non-negative integers n,

$$(-\infty) + (-\infty) = -\infty,$$

$-\infty < n$ for all non-negative integers n.

If the coefficient ring R may have proper divisors of zero, then the degree formula must be weakened to

$$\deg(f(x)g(x)) \leq \deg f(x) + \deg g(x).$$

A polynomial in two variables x, y, with coefficients in a ring R, can be regarded as a polynomial in y whose coefficients are polynomials in x, and conversely, i.e. we can write

$$R[x, y] = (R[x])[y].$$

And we can extend this to polynomials in any number of variables. It follows from Theorem 1.5.2 that, if R is an integral domain and $x_1, \ldots, x_n$ are independent variables, then the polynomial ring $R[x_1, \ldots, x_n]$ is also an integral domain.

It is an obvious remark that every subring of an integral domain is an integral domain. Thus, for example, the rings $\mathbb{Z}[\sqrt{5}]$ and $\mathbb{Z}[i]$ introduced in Section 1.4 are integral domains, because they are subrings of the integral domains $\mathbb{R}, \mathbb{C}$, respectively.

Exercises 1.5

1. Do the examples of rings in Exercises 1.3, 1.4 have proper divisors of zero?
2. Let R be a ring. We define the *order* of the non-zero power series $a_0 + a_1 x + a_2 x^2 + \ldots$ in $R[[x]]$ to be the smallest non-negative integer n such that $a_n \neq 0$. Write down a formula for $\operatorname{ord}(f(x)g(x))$, where $f(x), g(x) \in R[[x]]$ (a) when R has no proper divisors of zero, (b) when R may have proper divisors of zero. (You will need to define the order of the zero power series conventionally as ∞ and adopt conventions concerning the use of ∞.) Show that, when R is an integral domain, so is the power-series ring $R[[x]]$.

1.6 Fields

So far, the ring $\mathbb{Z}$ and polynomial rings have motivated our definitions. But there is one 'non-property' of these rings. It is not true that every non-zero integer has an integer inverse. Thus, for example, there is no integer n such that $2n = 1$, so that 2 has no inverse. Again, in a polynomial ring $R[x]$, where R is any ring, the polynomial x has no inverse. However, in $\mathbb{Q}$ 2 has inverse $\frac{1}{2}$, and in fact every non-zero rational number has a rational inverse, every non-zero real number has a real inverse and every non-zero complex number has a complex inverse.

We now make the following definitions.

Definition 1.6.1 Let R be a ring. An element u of R is said to be a 'unit' if it has an inverse, i.e. if there exists $u^{-1} \in R$ such that

$$u^{-1}u = uu^{-1} = 1.$$

It should be noted that each element u can have only one inverse u^{-1}. (To show this, adapt the argument in Definition 1.3.1 that an element can have only one negative.) Also, the product of two units u, v is again a unit, with $(uv)^{-1} = v^{-1}u^{-1}$, and the inverse of a unit u is a unit, with $(u^{-1})^{-1} = u$. Of course, 1 is a unit, with $1^{-1} = 1$. These remarks effectively say that the units of a ring R form a group under ring multiplication, called the *group of units* of R.

We next determine the units of a polynomial ring $R[x]$ when R is an integral domain. Suppose that $f(x)$ is a unit of $R[x]$. Then it has an inverse $g(x)$ in $R[x]$, so that

$$f(x)g(x) = 1.$$

Thus, because R is an integral domain,

$$\deg f(x) + \deg g(x) = \deg(f(x)g(x)) = 0,$$

so that $\deg f(x) = 0 = \deg g(x)$. Hence, $f(x) = u + 0x + 0x^2 + \ldots$, $g(x) = v + 0x + 0x^2 + \ldots$, and $uv = 1$. Thus u is a unit in R. Conversely, if u is a unit of R, then the polynomial $u + 0x + 0x^2 + \ldots$ is a unit of $R[x]$, with inverse

$u^{-1}+0x+0x^2+\ldots$. Thus the units of $R[x]$ are the polynomials of the form $u+0x+0x^2+\ldots$, where u is a unit of R. We always identity $u \in R$ with $u+0x+0x^2+\ldots \in R[x]$, in which case we can say that the units of $R[x]$ and the units of R are the same when R is an integral domain. This may not be true when R has proper divisors of zero. If we anticipate Section 1.7, we see that, with coefficients in the ring $\mathbb{Z}_4$ of integers modulo 4,

$$(\bar{1}+\bar{2}x)(\bar{1}+\bar{2}x) = \bar{1}+\bar{4}x+\bar{4}x^2 = \bar{1},$$

so that $\bar{1}+\bar{2}x$ is a unit in $\mathbb{Z}_4[x]$.

Definition 1.6.2 A 'field' is a commutative ring in which every non-zero element is a unit.

Thus $\mathbb{Q}$, $\mathbb{R}$ and $\mathbb{C}$ are examples of fields.

We shall shortly provide ways of constructing fields. But first we consider the connection between integral domains and fields.

Theorem 1.6.3 Every field is an integral domain.

Proof Let F be a field. We have to show that F has no proper divisors of zero. Let $a, b \in F$ be such that $ab=0$, and suppose that $a \neq 0$. Then a is a unit, so we can write $a^{-1}ab = a^{-1}0 = 0$, i.e. $b=0$, and F has no proper divisors of zero. □

The integral domain $\mathbb{Z}$ tells us that not every integral domain is a field. However, we have the following partial converse of Theorem 1.6.3.

Theorem 1.6.4 Every finite integral domain is a field.

Proof Let R be a finite integral domain, and denote its distinct elements by $a_1, a_2, \ldots, a_n$. Consider a non-zero element a of R. We must show that a is a unit. Consider the elements $aa_1, aa_2, \ldots, aa_n$. No two of these can be the same, because if $i \neq j$ then $a_i \neq a_j$ so $aa_i \neq aa_j$ by the cancellation law. Thus $aa_1, aa_2, \ldots, aa_n$ are again the elements of R in some order. In particular, there exists i such that $aa_i = 1$, and a has inverse a_i and so is a unit. □

1.6 Fields

Although $\mathbb{Z}$ is not a field, it has a field closely associated with it, namely the field $\mathbb{Q}$ of rational numbers. In fact, a rational number is just a fraction formed from two integers, its numerator and its denominator. The construction of rational numbers from integers can be applied to any integral domain. Thus, we begin with an arbitrary integral domain R, and consider the set S of all ordered pairs (a, b), where $a, b \in R$ and $b \neq 0$. However, a rational number is not just a pair of integers, because two different pairs may give the same rational number (e.g. $\frac{1}{2} = \frac{2}{4}$). We therefore introduce a relation $\sim$ on S by defining

$$(a, b) \sim (c, d) \Leftrightarrow ad = bc.$$

This relation is easily seen to be an equivalence relation. (1) It is reflexive because $ab = ba$ so $(a, b) \sim (a, b)$. (2) It is symmetric because $(a, b) \sim (c, d) \Rightarrow ad = bc \Rightarrow cb = da \Rightarrow (c, d) \sim (a, b)$. (3) It is transitive because $(a, b) \sim (c, d)$ and $(c, d) \sim (e, f) \Rightarrow ad = bc$ and $cf = de$, so that $adf = bcf = bde$, and we can cancel d because $d \neq 0$ to give $af = be$, i.e. $(a, b) \sim (e, f)$.

An equivalence relation on a set has the effect of partitioning the set into disjoint subsets. In our case, we denote the subset of this partition to which (a, b) belongs by a/b. Thus a fraction is a set in the partition, although it is much the best to regard it as a single element satisfying the identity

$$\frac{a}{b} = \frac{c}{d} \Leftrightarrow ad = bc \ (b, d \neq 0).$$

We denote the set of all these fractions by k.

We now state the rules for adding and multiplying fractions. They are:

$$\frac{a}{b} + \frac{c}{d} = \frac{ad + bc}{bd}, \quad \frac{a}{b} \cdot \frac{c}{d} = \frac{ac}{bd}.$$

Care needs to be taken at this point. Because of the ambiguity in denoting a given fraction, we need to verify that we have well-defined binary operations. Suppose that

$$\frac{a}{b} = \frac{a_1}{b_1}, \quad \frac{c}{d} = \frac{c_1}{d_1}.$$

Then $ab_1 = ba_1$, $cd_1 = dc_1$, so

$$(ad+bc)b_1d_1 = (ab_1)dd_1 + (cd_1)bb_1 = (ba_1)dd_1 + (dc_1)bb_1$$
$$= bd(a_1d_1 + b_1c_1),$$

and

$$(ac)(b_1d_1) = (ab_1)(cd_1) = (ba_1)(dc_1) = (bd)(a_1c_1),$$

so that

$$\frac{ad+bc}{bd} = \frac{a_1d_1 + b_1c_1}{b_1d_1}, \quad \frac{ac}{bd} = \frac{a_1c_1}{b_1d_1}.$$

Thus the addition and the multiplication on k are well-defined. It is an elementary but tedious matter to verify that the axioms for a field are satisfied by k. We note that the zero element is $0/1$ and the identity element is $1/1$; also that, when a/b is not zero, it has inverse b/a.

There is a natural way of embedding R in k. We define the mapping $R \to k$ by $a \mapsto a/1$. It is easy to see that this is an embedding. Thus R is isomorphic to a subring of k. It is usual to regard the fraction $a/1$ and a as one and the same (so that, in $\mathbb{Z}$ and $\mathbb{Q}$, the integer 2 and the rational number $2/1$ are identified), when this embedding becomes inclusion and R is a subring of k. We call k the *field of fractions* of R.

Exercises 1.6

1. Show that $\mathbb{Q}[\sqrt{2}]$, consisting of all real numbers of the form $a_0 + a_1\sqrt{2} + a_2(\sqrt{2})^2 + \ldots$, where $a_0, a_1, a_2, \ldots \in \mathbb{Q}$ with only finitely many a_i non-zero (or, equivalently, $\mathbb{Q}[\sqrt{2}] = \{a + b\sqrt{2} : a, b \in \mathbb{Q}\}$), is a field under the addition and multiplication of real numbers.

2. Denote by I, A, B, C the 2×2 matrices

$$\begin{bmatrix} 1 & 0 \\ 0 & 1 \end{bmatrix}, \begin{bmatrix} 0 & 1 \\ -1 & 0 \end{bmatrix}, \begin{bmatrix} 0 & i \\ i & 0 \end{bmatrix}, \begin{bmatrix} i & 0 \\ 0 & -i \end{bmatrix}$$

with complex entries, respectively, and put

$$\mathbb{H} = \{aI + bA + cB + dC : a, b, c, d \in \mathbb{R}\}.$$

1.6 Fields

Show that

$$A^2 = B^2 = C^2 = -I, \ BC = -CB = A,$$
$$CA = -AC = B, \ AB = -BA = C.$$

Deduce that $\mathbb{H}$ is a subring of the matrix ring $M_2(\mathbb{C})$. ($\mathbb{H}$ is called the *ring of quaternions*.) Let

$$x = aI + bA + cB + dC \in \mathbb{H},$$

where $a, b, c, d \in \mathbb{R}$, be a non-zero quaternion, and write

$$\bar{x} = aI - bA - cB - dC.$$

Compute $x\bar{x}$ and $\bar{x}x$, and deduce that every non-zero quaternion is a unit. Does this make $\mathbb{H}$ into a field? How may $\mathbb{C}$ be embedded in $\mathbb{H}$? (The quaternion $aI + bA + cB + dC$ as given here is usually rewritten as $a + bi + cj + dk$, with $i^2 = j^2 = k^2 = -1$, $jk = -kj = i$, $ki = -ik = j$, $ij = -ji = k$. The idea of these 'numbers' came to the distinguished Irish mathematician Sir William Hamilton in 1843 as a tool in his work on three-dimensional dynamics–he needed four-dimensional numbers! Since then, the a has been dropped to give the now-familiar three-dimensional vector $bi + cj + dk$, and $jk = -kj = i$ has become the vector-product formula $j \wedge k = -k \wedge j = i$ etc.)

3. Let R be a commutative ring. Show that, in the power series ring $R[[x]]$, $a_0 + a_1 x + a_2 x^2 + \ldots$ (with the a_i's in R) is a unit if and only if a_0 is a unit of R. What is the inverse of $1 + x$ in $\mathbb{Z}[[x]]$? Deduce that every non-zero power series in $k[[x]]$, where k is a field, can be expressed in the form $x^n u$, where u is a unit in $k[[x]]$.

4. Denote by F the set of all matrices of the form

$$\begin{bmatrix} a & b \\ -b & a \end{bmatrix},$$

where $a, b \in \mathbb{R}$. Show that F is a field under matrix addition and multiplication.

28   Rings

5. Show that the cartesian product $\mathbb{R} \times \mathbb{R}$ with addition and multiplication defined by

$$(a, b) + (c, d) = (a+c, b+d),$$
$$(a, b)(c, d) = (ac - bd, ad + bc)$$

is a field. Which field is this?

6. Show that the matrices of the form

$$\begin{bmatrix} a & 2b \\ b & a \end{bmatrix},$$

where a, b are rational numbers, form a field under matrix addition and multiplication.

7. Let R be a finite integral domain. By considering powers of a non-zero element a of R, provide another proof of Theorem 1.6.4, showing also that the inverse of a is a non-negative power of a. Looking ahead to Section 1.7, find the smallest such power for each non-zero element of $\mathbb{Z}_7$.

1.7

Residue-class rings

In Theorem 1.6.4 we proved that every finite integral domain is a field, but gave no indication of where such things might be found. We shall rectify that in this section.

We begin with the set $\mathbb{Z}$ of integers and a fixed integer $n > 1$. We recall that integers a, b are said to be *congruent modulo n*, and we write

$$a \equiv b \pmod{n},$$

if the difference between a and b is divisible by n; or, in symbols, if $n|(a-b)$. Thus, for example, $-5 \equiv 11 \pmod{4}$ because -16 is divisible by 4. Congruence modulo n is easily seen to be an equivalence relation. (1) It is reflexive because $a \equiv a \pmod{n}$ for all $a \in \mathbb{Z}$. (2) It is symmetric because if $a \equiv b \pmod{n}$ then $b \equiv a \pmod{n}$. (3) It is transitive because if $a \equiv b \pmod{n}$ and $b \equiv c \pmod{n}$ then $a \equiv c \pmod{n}$. Thus congruence modulo n partitions $\mathbb{Z}$ into disjoint

1.7 Residue-class rings

subsets. We denote the subset to which a belongs by $\bar{a}$, and call these the residue classes of integers modulo n. Thus, for $a, b \in \mathbb{Z}$,

$$\bar{a} = \bar{b} \Leftrightarrow a \equiv b \pmod{n}. \tag{*}$$

We denote by $\mathbb{Z}_n$ the set of all residue classes of integers modulo n. In fact, $\mathbb{Z}_n$ has precisely n distinct elements, namely $\bar{0}, \bar{1}, \bar{2}, \ldots, \overline{n-1}$. For example, $\mathbb{Z}_4$ has four distinct elements $\bar{0}, \bar{1}, \bar{2}, \bar{3}$ with $\bar{4} = \bar{0}$, $\bar{5} = \bar{1}$, $\overline{-1} = \bar{3}$ etc. It is much the best to regard $\bar{a}$ as a single element of $\mathbb{Z}_n$ rather than as a set of elements of $\mathbb{Z}$, with $\bar{a}$'s satisfying the rule (*) above.

We now attempt to introduce an addition and multiplication on $\mathbb{Z}_n$. We write

$$\bar{a} + \bar{b} = \overline{a+b},\ \bar{a}\bar{b} = \overline{ab}\ (a, b \in \mathbb{Z}).$$

Thus, for example, in $\mathbb{Z}_4$, $\bar{2} + \bar{3} = \bar{5} = \bar{1}$, $\bar{2} \cdot \bar{3} = \bar{6} = \bar{2}$. Because of the ambiguity of writing down elements of $\mathbb{Z}_n$ as given by (*), we need to verify that these rules give well-defined binary operations on $\mathbb{Z}_n$. Suppose that $\bar{a} = \bar{a}_1$, $\bar{b} = \bar{b}_1$, where $a, b, a_1, b_1 \in \mathbb{Z}$. Then $a \equiv a_1 \pmod{n}$, $b \equiv b_1 \pmod{n}$, so $a = a_1 + kn, b = b_1 + ln$ for some $k, l \in \mathbb{Z}$. Then

$$a+b = (a_1 + b_1) + (k+l)n,$$
$$ab = a_1 b_1 + (kb_1 + la_1 + kln)n,$$

so that

$$a+b \equiv a_1 + b_1 \pmod{n},\ ab \equiv a_1 b_1 \pmod{n}$$

and $\overline{a+b} = \overline{a_1 + b_1}, \overline{ab} = \overline{a_1 b_1}$. This verifies that addition and multiplication are well-defined on $\mathbb{Z}_n$. It is now an elementary if tedious matter to verify that $\mathbb{Z}_n$ is a commutative ring. Its zero element is $\bar{0}$, its identity element is $\bar{1}$ and $\bar{1} \neq \bar{0}$ because $n > 1$, and $-\bar{a} = \overline{-a}$.

Many of these residue-class rings have proper divisor of zero, and so are not integral domains and therefore are not fields. For example, in $\mathbb{Z}_4$ $\bar{2} \cdot \bar{2} = \bar{4} = \bar{0}$, and in $\mathbb{Z}_{12}$ $\bar{2} \cdot \bar{6} = \overline{12} = \bar{0}$. We first work out which elements of $\mathbb{Z}_n$ are units.

Theorem 1.7.1 Let $a \in \mathbb{Z}$. The element $\bar{a}$ of $\mathbb{Z}_n$ is a unit if and only if a, n are coprime (i.e. they have greatest common divisor 1).

Example Thus, in $\mathbb{Z}_{10}$, the units are $\bar{1}, \bar{3}, \bar{7}, \bar{9}$. In fact, $\bar{1}^{-1} = \bar{1}, \bar{3}^{-1} = \bar{7}, \bar{7}^{-1} = \bar{3}, \bar{9}^{-1} = \bar{9}$.

Remark Readers unfamiliar with the theory of highest common factors of integers may be advised to return to Theorem 1.7.1 when this has been developed in Part 2. We promise readers that Theorem 1.7.1 is not used to prove results in Part 2, thereby bringing in the danger of a circular argument. This section is included here primarily to provide more examples of rings and in particular of finite fields.

Proof of Theorem 1.7.1 First assume that $\bar{a}$ is a unit in $\mathbb{Z}_n$, so that there exists $s \in \mathbb{Z}$ such that $\bar{s}\bar{a} = \bar{1}$. Then $\overline{sa} = \bar{1}$, so that

$$sa \equiv 1 \pmod{n},$$

whence there exists $t \in \mathbb{Z}$ such that

$$sa - 1 = tn.$$

It follows that any integer dividing a and n must also divide 1. Certainly 1 is a common divisor of a, n. It follows that 1 is a greatest common divisor of a, n and a, n are thereby coprime.

Now assume that a, n are coprime. This means that their greatest common divisor is 1, so there exist $s, t \in \mathbb{Z}$ such that $sa + tn = 1$ (see Theorem 2.4.2). Now

$$\overline{sa + tn} = \bar{1},$$

whence

$$\bar{s}\bar{a} + \bar{t}\bar{n} = \bar{1}.$$

But $\bar{n} = \bar{0}$ in $\mathbb{Z}_n$, so that $\bar{s}\bar{a} = \bar{1}$ and $\bar{a}$ is a unit in $\mathbb{Z}_n$, with inverse $\bar{s}$. □

This means that every non-zero element of $\mathbb{Z}_n$ is a unit if and only if $1, 2, \ldots, n-1$ are all coprime to n, i.e. if and only if n is prime. This gives the following corollary:

Corollary 1.7.2 $\mathbb{Z}_p$ is a field if and only if p is prime. □

This determines fields of each prime order. (The 'order' of a field is the numbers of its elements.) It is not difficult to see from the theory

1.7 Residue-class rings

of dimension in vector spaces that every finite field must have order p^n for some prime number p and positive integer n. The brilliant French mathematician Evariste Galois (1811–32) showed how to construct a finite field of order p^n for every prime number p and every positive integer n, now called *the Galois field of order p^n*. Theorem 1.7.1 and Corollary 1.7.2 suggest a connection between abstract algebra and number theory at this point which is worth developing further. This provides some justification for the abstract algebra and also highlights two beautiful and important results in elementary number theory.

Theorem 1.7.3 (Fermat's Little Theorem) Let p be a prime number and let n be an integer not divisible by p. Then

$$n^{p-1} \equiv 1 \pmod{p}.$$

Further, for all integers n,

$$n^p \equiv n \pmod{p}.$$

Example Put $p=7$, $n=5$. Then $5^6 \equiv (-2)^6 \equiv 64 \equiv 1 \pmod{7}$.

Historical note Pierre Fermat was born in 1601 in France. His father was a leather merchant. He spent most of his time as a civil servant. Along with Descartes he helped to found coordinate geometry, and he was also one of the originators of calculus. He is regarded, too, as the father of probability theory. He claimed his Little Theorem in a letter he wrote in 1640, but the first published proof was by Gauss in his *Disquisitiones Arithmeticae*. Fermat is reported to have said on one occasion, 'I think I see a great light'.

Proof of Fermat's Little Theorem We first suppose that p does not divide n. The non-zero elements of the field $\mathbb{Z}_p$ are $\bar{1}, \bar{2}, \ldots, \overline{p-1}$. Consider the elements

$$\bar{1} \cdot \bar{n},\ \bar{2} \cdot \bar{n},\ \bar{3} \cdot \bar{n}, \ldots, \overline{p-1} \cdot \bar{n}$$

of $\mathbb{Z}_p$. Since $\bar{n} \neq \bar{0}$ and the cancellation law holds in $\mathbb{Z}_p$, these are all distinct. They are also non-zero because $\mathbb{Z}_p$ has no proper divisors of zero. Hence, these are all the non-zero elements of $\mathbb{Z}_p$ back again in some order, so that

$$(\bar{1}\cdot\bar{n})(\bar{2}\cdot n)\ldots(\overline{p-1}\cdot\bar{n})=\overline{1\cdot 2\cdot\ldots\cdot p-1}.$$

We can now cancel $\overline{1\cdot 2\cdot\ldots\cdot p-1}$ to give

$$\bar{n}^{p-1}=\bar{1},$$

or

$$\overline{n^{p-1}}=\bar{1},$$

so that

$$n^{p-1}\equiv 1 \text{ (mod } p).$$

We can multiply by n to give

$$n^p\equiv n \text{ (mod } p),$$

and this is also true when n is divisible by p, because then both sides of the congruence are congruent to zero. □

For those who are familiar with Lagrange's Theorem for finite groups, Fermat's Little Theorem can be deduced very easily from Lagrange's Theorem. The non-zero elements of the field $\mathbb{Z}_p$ form a group of order $p-1$, its group of units. If n is not divisible by p, then $\bar{n}$ is an element of this group and it follows easily from Lagrange's Theorem that the order of the group element $\bar{n}$ divides $p-1$, whence $\bar{n}^{p-1}=\bar{1}$.

Theorem 1.7.4 (Wilson's Theorem) Let p be a prime number. Then

$$(p-1)!\equiv -1 \text{ (mod } p).$$

Example Put $p=11$. Then

$$10!=10\times 9\times 8\times 7\times 6\times 5\times 4\times 3\times 2\times 1$$
$$\equiv(-1)\times(-2)\times(-3)\times(-4)\times(-5)\times 5\times 4$$
$$\quad\times 3\times 2\times 1 \text{ (mod 11)}$$
$$\equiv -1^2\times 2^2\times 3^2\times 4^2\times 5^2 \text{(mod 11)}$$
$$\equiv -10^2\times 12^2 \text{ (mod 11)}$$
$$\equiv -(-1)^2\times 1^2 \text{ (mod 11)}$$
$$\equiv -1 \text{ (mod 11)}.$$

1.7 Residue-class rings

Historical note John (later Sir John) Wilson (1741–93) was a lawyer and a judge. He was a pupil of Waring. This result was ascribed to him in 1770, although Leibniz probably knew the result 100 years before. The first published proof was by Lagrange in 1773. A reported 'proof' of Wilson's Theorem by a Spanish nobleman goes as follows:

$$(p-1)+1 = p$$

so

$$(p-1)! + 1! = p!$$

so

$$(p-1)! \equiv -1 \pmod{p}.$$

We leave readers to discover the flaw!

Proof of Wilson's Theorem First consider an arbitrary field F. If $a \in F$ is its own inverse, then

$$a^2 = 1$$
$$\Rightarrow a^2 - 1 = 0$$
$$\Rightarrow (a-1)(a+1) = 0$$
$$\Rightarrow a-1 = 0 \text{ or } a+1 = 0$$
$$\Rightarrow a = 1 \text{ or } a = -1.$$

Certainly $1, -1$ are their own inverses, so F has at most two elements which are their own inverses ('at most' because it could be true that $1 = -1$ in F–an example of this would be in the field $\mathbb{Z}_2$).

Now let F be a finite field. The elements which are not their own inverses can be paired off with their inverses. This leaves $1, -1$, which could be equal. Thus, when we multiply together all the non-zero elements of F, we obtain a product of the form

$$x_1 \times x_1^{-1} \times x_2 \times x_2^{-1} \times \cdots \times x_r \times x_r^{-1} \times 1 \times (-1)$$

if $1 \neq -1$, which is equal to -1. If $1 = -1$, the product is of the form

$$x_1 \times x_1^{-1} \times x_2 \times x_2^{-1} \times \cdots \times x_r \times x_r^{-1} \times (-1),$$

and again is -1 (or 1). Thus the product of the non-zero elements of a finite field is always -1.

We now apply this remark to the field $\mathbb{Z}_p$, to give

$$\bar{1} \cdot \bar{2} \cdot \ldots \cdot \overline{p-1} = -\bar{1}$$
$$\Rightarrow \overline{(p-1)!} = \overline{-1}$$
$$\Rightarrow (p-1)! \equiv -1 \pmod{p}. \quad \square$$

Exercises 1.7
1. Find the remainder when 3^{47} is divided by 23.
2. Find integers $n, k > 1$ such that $n^k \not\equiv n \pmod{k}$.
3. Show that $n^{561} \equiv n \pmod{561}$ for all $n \in \mathbb{Z}$.
4. Show that $143^6 + 91^{10} + 77^{12} \equiv 1 \pmod{1001}$.
5. Show that, for every positive integer n, $n^{37} - n$ is divisible by 383838.
6. Show that $2222^{5555} + 5555^{2222}$ is divisible by 7.
7. Prove the converse of Wilson's Theorem, namely that, if n is an integer greater than 1 such that $(n-1)! \equiv -1 \pmod{n}$, then n must be prime. Use this to show that 17 is prime.
8. Let m, n be positive integers with $n > 1$, and define

$$k = (n-1)! + 1 - mn,$$
$$p = \tfrac{1}{2}(n-2)\{|k^2 - 1| - (k^2 - 1)\} + 2.$$

Show that p takes all prime values and no others as m, n vary. Find m, n which make $p = 7$. (This was pointed out in an article in *The Guardian* newspaper by Keith Devlin, and gives the lie to the oft-repeated assertion – of which the present author is guilty! – that there is no known function which takes all prime values and no others. Not that this function is of any use in generating prime numbers in practice!)
9. Let p be a prime number. Show that p divides $n^p + (p-1)!n$ for all integers n.
10. Let p be an odd prime number. Show that

$$1^2 \times 3^2 \times 5^2 \times \ldots \times (p-2)^2 \equiv (-1)^{(p+1)/2} \pmod{p}.$$

Part 2

Factorization

2.1

Introduction

We are now ready to develop the theory of factorization in a more general setting than with integers or polynomials alone. There is one complication when dealing with factorization in general which can largely be avoided when dealing with integers or polynomials. When factorizing integers, we can stick to positive integers and positive factors and everything is unique, so that $60 = 2 \times 2 \times 3 \times 5$ and there is no other way of factorizing 60 into prime factors, apart of course from changing the order of the factors. Similarly, with polynomials we can stick to 'monic polynomials', i.e. polynomials whose 'leading coefficient' – the coefficient of the term of highest degree – is 1. Thus, in $\mathbb{R}[x]$,

$$x^3 + x^2 + x + 1 = (x+1)(x^2+1),$$

and there is no other way of factorizing this polynomial into monic irreducible polynomials. If we drop the insistence on positive integers, then of course we can write $60 = (-2) \times (-2) \times 3 \times 5$, for instance, and change the factorization. Similarly, if we drop the insistence that polynomial factors are monic, we can write

$$x^3 + x^2 + x + 1 = (2x+2)(\tfrac{1}{2}x^2 + \tfrac{1}{2}).$$

The ambiguity arises from the units of the two rings. The units of $\mathbb{Z}$ are 1, -1, and the units of $\mathbb{R}[x]$ are the non-zero constant polynomials. In general, a given factorization can always be changed by scattering units around. Also, if you are trying to factorize an element of a ring into two factors, and you can only do it by having one of the factors a unit, then you have cheated and the element you began with cannot really be factorized.

The appropriate setting when dealing with factorization in general, at least at our modest level, is an integral domain. Thus the symbol R will denote an integral domain throughout Part 2, unless otherwise stated.

2.2

Unique factorization domains

We first make precise the notion of divisibility.

Definition 2.2.1 Let $a, b \in R$. We say that 'a is a factor of b', or 'a is a divisor of b' or 'a divides b', and write $a|b$, if there exists $c \in R$ such that $b = ac$.

Notes (1) Divisibility is a reflexive and transitive relation on R. Thus $a|a$ for all $a \in R$ because $a = 1a$, and $a|b, b|c \Rightarrow a|c$ for $a, b, c \in R$. But $a|b \not\Rightarrow b|a$ in general, so it is not symmetric.

(2) Let $u \in R$. Then u is a unit if and only if $u|1$.

(3) It follows from (2) and the fact that divisibility is transitive that every factor of a unit is again a unit.

Definition 2.2.2 Let $a, b \in R$. We say that 'a is an associate of b', and write $a \sim b$, if $a = ub$ for some unit u of R.

Notes (1) $\sim$ is an equivalence relation, so we can refer simply to 'associate elements a, b'.

(2) If a, b are associates, then $a|b$ and $b|a$. The converse is also true. For suppose that $a|b$ and $b|a$. Then there exist $u, v \in R$ such that $b = ua, a = vb$, so $b = uvb$. If $b \neq 0$, then we can cancel b to give $1 = uv$ and u is a unit, so that $a \sim b$. If $b = 0$, then also $a = 0$ and again $a \sim b$.

(3) If u is a unit of R and $a \in R$, then $u|a$ and $ua|a$ because $a = u(u^{-1}a) = u^{-1}(ua)$. Factors of a other than units and associates of a are said to be *proper factors* of a. Thus $\pm 2, \pm 3$ are proper factors of 6 in $\mathbb{Z}$, but 7 has no proper factors.

(4) Suppose that $a = bc$ with $a, b, c \in R$. If b is a unit, then c will be an associate of a. If c is an associate of a and $a \neq 0$, then b will be a unit.

2.2 Unique factorization domains

Definition 2.2.3 An element p of R is said to be an 'irreducible element' of R if (1) $p \neq 0$, (2) p is not a unit, and (3) p has no proper factors, i.e. $p = ab \Rightarrow a$ or b is a unit (or, equivalently, $a \sim p$ or $b \sim p$).

Notes (1) The positive integers satisfying the properties of this definition are usually referred to as the 'prime numbers'. The word 'prime' is reserved here for another type of element which is the same as an irreducible element in the case of the integers. More of this later.

(2) If p is irreducible, then it is easy to see that so is up, where u is a unit.

Remark Let $a \in R$, and suppose that there are irreducible elements $p_1, p_2, \ldots, p_s$ such that

$$a = p_1 p_2 \cdots p_s.$$

We call such an expression a factorization of a into irreducible elements. If we have units $u_1, u_2, \ldots, u_s$ such that $u_1 u_2 \cdots u_s = 1$ (and we can always obtain such by choosing units $u_1, u_2, \ldots, u_{s-1}$ at will and writing $u_s = u_1^{-1} u_2^{-1} \cdots u_{s-1}^{-1}$), then a also has the factorization

$$a = (u_1 p_1)(u_2 p_2) \cdots (u_s p_s)$$

into irreducible elements. To this extent, factorization into irreducibles can never be unique, as we pointed out in the introductory remarks in Section 2.1. Note that $p_1 \sim u_1 p_1, \ldots, p_s \sim u_s p_s$.

We now come to the crucial definition.

Definition 2.2.4 We say that R is a 'unique factorization domain' (UFD for short) if (1) every element of R which is neither zero nor a unit can be expressed in the form $p_1 p_2 \cdots p_s$, where the p_i are irreducibles, and (2) whenever

$$p_1 p_2 \cdots p_s = q_1 q_2 \cdots q_t,$$

where the p_i, q_j are irreducibles, then $s = t$ and we can rearrange the q_j's in the product such that $p_i \sim q_i$ for $1 \leq i \leq s$.

The prototype examples of UFDs are $\mathbb{Z}$ (from the Fundamental

Theorem of Arithmetic) and $\mathbb{C}[x]$ and $\mathbb{R}[x]$ (from the Fundamental Theorem of Algebra). In fact, $k[x]$ is a UFD for every field k. That these are examples of UFDs will emerge from the more general considerations which follow.

At this stage, readers may be intrigued to see an example of an integral domain which is not a UFD. We consider the set $\mathbb{Z}[\sqrt{-5}]$ of all complex numbers of the form

$$a_0 + a_1\sqrt{-5} + a_2(\sqrt{-5})^2 + \ldots,$$

where the a_i are integers. This simplifies to the set of all complex numbers of the form $a + b\sqrt{-5}$, where $a, b \in \mathbb{Z}$. As we indicated in Section 1.4, this is a subring of $\mathbb{C}$. Since $\mathbb{C}$ is an integral domain, $\mathbb{Z}[\sqrt{-5}]$ is thus an integral domain under the addition and multiplication of complex numbers. For $\alpha = a + b\sqrt{-5}$ ($a, b \in \mathbb{Z}$), we define the *norm* $N(\alpha)$ of α by

$$N(\alpha) = |\alpha|^2 = a^2 + 5b^2.$$

Thus $N(\alpha)$ is a non-negative integer. Also, if $\alpha, \beta \in \mathbb{Z}[\sqrt{-5}]$, then

$$N(\alpha\beta) = |\alpha\beta|^2 = |\alpha|^2|\beta|^2 = N(\alpha)N(\beta).$$

Let u be a unit of $\mathbb{Z}[\sqrt{-5}]$. Then $uu^{-1} = 1$, so

$$N(u)N(u^{-1}) = N(uu^{-1}) = N(1) = 1.$$

Since $N(u)$ is a non-negative integer, this means that $N(u) = 1$. Thus every unit has norm 1. Consider an element $u = a + b\sqrt{-5}$ ($a, b \in \mathbb{Z}$) of norm 1. Then $a^2 + 5b^2 = 1$, so $a = \pm 1, b = 0$. Thus the only elements of norm 1 are ± 1, and these are units. Thus the units of $\mathbb{Z}[\sqrt{-5}]$ are just the elements of norm 1 and are precisely $1, -1$. Now, in $\mathbb{Z}[\sqrt{-5}]$,

$$9 = 3 \times 3 = (2 + \sqrt{-5})(2 - \sqrt{-5}).$$

We assert that $3, 2 + \sqrt{-5}, 2 - \sqrt{-5}$ are irreducible. They are all of norm 9. Let α be an element of $\mathbb{Z}[\sqrt{-5}]$ of norm 9, and write $\alpha = \beta\gamma$, where $\beta, \gamma \in \mathbb{Z}[\sqrt{-5}]$. Then

$$9 = N(\alpha) = N(\beta\gamma) = N(\beta)N(\gamma).$$

It is easy to see that no element of $\mathbb{Z}[\sqrt{-5}]$ has norm 3, so either $N(\beta) = 1$ or $N(\gamma) = 1$. Thus either β or γ is a unit. Thus an element of

2.3 Euclidean domains

norm 9 is irreducible. Since $3 \nmid 2 \pm \sqrt{-5}$, we thus have two essentially different factorizations of 9 into irreducibles, so $\mathbb{Z}[\sqrt{-5}]$ is not a UFD.

We shall see later that $\mathbb{Z}[i]$, $\mathbb{Z}[\sqrt{-2}]$ are UFDs. This will be part of a bigger search for a class of rings which are UFDs. The concept of unique factorization goes back to Euclid, and it is to Euclid that we turn for such a class of rings, appropriately termed 'Euclidean domains'.

Exercises 2.2

1. Provide two essentially different factorizations of 6 into irreducibles in $\mathbb{Z}[\sqrt{-5}]$.
2. For $\alpha = a + b\sqrt{5} \in \mathbb{Z}[\sqrt{5}]$, where $a, b \in \mathbb{Z}$, put $N(\alpha) = a^2 - 5b^2$.

    (a) Show that $N(\alpha\beta) = N(\alpha)N(\beta)$ for all $\alpha, \beta \in \mathbb{Z}[\sqrt{5}]$.

    (b) Show that α is a unit if and only if $N(\alpha) = \pm 1$, and give an example of a unit of norm 1 other than ± 1 and a unit of norm -1.

    (c) If $N(\alpha)$ is a prime number, show that α is irreducible.

    (d) Show that the congruences $x^2 \equiv \pm 2 \pmod{5}$ have no solutions in $\mathbb{Z}$, and deduce that 2, $3 + \sqrt{5}$, $3 - \sqrt{5}$ are irreducible in $\mathbb{Z}[\sqrt{5}]$.

    (e) Show that $4 = 2 \times 2 = (3 + \sqrt{5})(3 - \sqrt{5})$ gives two essentially different factorizations of 4 into irreducibles in $\mathbb{Z}[\sqrt{5}]$, so that $\mathbb{Z}[\sqrt{5}]$ is not a UFD.

2.3

Euclidean domains

Consider the ring $\mathbb{Z}$. This ring is not a field, so it is not usually possible to divide one integer by another and obtain an integer. But there is a concept of division in $\mathbb{Z}$ which goes back to Euclid's *Elements*. Suppose, for example, that we try to divide 13 by 5. We say that the quotient is 2 and the remainder is 3, because

$$13 = 2 \times 5 + 3.$$

This happens generally, as the next theorem shows.

Theorem 2.3.1 (The Division Algorithm for Integers) Let $m, n \in \mathbb{Z}$ with $n \neq 0$. Then there exist unique integers q, r such that

$$m = qn + r \quad \text{and} \quad 0 \leq r < |n|.$$

In this case we say that 'q is the quotient and r the remainder when m is divided by n'.

Proof Consider the set S consisting of all non-negative integers of the form $m - xn$, where $x \in \mathbb{Z}$. Then S is not empty because if $m \geq 0$ then $m = m - 0 \times n \in S$ and if $m < 0$ then $m(1 - n^2) = m - (mn)n \in S$. It follows from the Well-Ordering Principle† that S has a smallest member r (say). Thus $r = m - qn$ for some $q \in \mathbb{Z}$. Now $r < |n|$ because if $r \geq |n|$ then

$$r - |n| = m - (q \pm 1)n \in S$$

yet is smaller than the smallest member of S. Thus we have found integers q, r with the required properties.

To prove uniqueness, suppose that $q_1, q_2, r_1, r_2 \in \mathbb{Z}$ are such that

$$q_1 n + r_1 = q_2 n + r_2 \quad \text{with} \quad 0 \leq r_1, r_2 < |n|.$$

Suppose that $q_1 \neq q_2$. Then

$$|r_1 - r_2| = |q_2 - q_1||n| \geq |n|,$$

which is impossible. Hence, $q_1 = q_2$, and so also $r_1 = r_2$. □

A similar thing happens with polynomials, as the next theorem shows.

Theorem 2.3.2 (The Division Algorithm for Polynomials) Let k be a field and let $f(x), g(x) \in k[x]$, where $g(x)$ is not the zero polynomial. Then there exist unique polynomials $q(x), r(x) \in k[x]$ such that

$$f(x) = q(x)g(x) + r(x) \quad \text{and} \quad \deg r(x) < \deg g(x).$$

Again we say that '$q(x)$ is the quotient and $r(x)$ the remainder when $f(x)$ is divided by $g(x)$'.

---

† The Well-Ordering Principle says that every non-empty set of non-negative integers has a least member. It is equivalent to the Principle of Induction, one of Peano's axioms for the integers.

2.3 Euclidean domains

Proof The proof follows similar lines to that of the division algorithm for integers. Denote by S the set of all polynomials of the form $f(x) - h(x)g(x)$, where $h(x) \in k[x]$. Then S is not empty. If the zero polynomial belongs to S, then there exists $h(x) \in k[x]$ such that $f(x) - h(x)g(x) = 0$, and we can take $q(x) = h(x)$ and $r(x)$ to be the zero polynomial with degree $-\infty < \deg g(x)$. Suppose now that the zero polynomial does not belong to S. Then the set of degrees of polynomials of S is a non-empty set of non-negative integers and so has a least member by the Well-Ordering Principle. Thus S has a polynomial $r(x)$ of least degree, and $r(x)$ can be expressed in the form

$$r(x) = f(x) - q(x)g(x),$$

for some $q(x) \in k[x]$. We claim that $\deg r(x) < \deg g(x)$. Suppose the contrary, and write

$$r(x) = a_m x^m + a_{m-1} x^{m-1} + \ldots + a_0,$$
$$g(x) = b_n x^n + b_{n-1} x^{n-1} + \ldots + b_0,$$

where the a_i's and b_j's belong to k and $a_m \neq 0, b_n \neq 0$. Then $m \geq n$. Now

$$r(x) - a_m b_n^{-1} x^{m-n} g(x)$$
$$= f(x) - (q(x) + a_m b_n^{-1} x^{m-n})g(x),$$

and so belongs to S. But

$$r(x) - a_m b_n^{-1} x^{m-n} g(x)$$
$$= (a_m x^m + \text{terms of lower degree})$$
$$- (a_m x^m + \text{terms of lower degree}),$$

so that $\deg(r(x) - a_m b_n^{-1} x^{m-n} g(x)) < m = \deg r(x)$ and S has a member whose degree is smaller than that of $r(x)$. This contradicts the choice of $r(x)$. Hence, $\deg r(x) < \deg g(x)$ after all, and the existence of $q(x), r(x)$ has been established.

To prove uniqueness, suppose that $q_1(x), q_2(x), r_1(x), r_2(x) \in k[x]$ are such that

$$q_1(x)g(x) + r_1(x) = q_2(x)g(x) + r_2(x)$$

with $\deg r_1(x) < \deg g(x)$, $\deg r_2(x) < \deg g(x)$. Suppose that

$q_1(x) \neq q_2(x)$. Then

$$\deg(r_1(x) - r_2(x)) = \deg((q_2(x) - q_1(x))g(x))$$
$$= \deg(q_2(x) - q_1(x)) + \deg g(x)$$
$$\geq \deg g(x).$$

But it is easy to see that

$$\deg(r_1(x) - r_2(x))$$
$$\leq \max \{\deg r_1(x), \deg r_2(x)\} < \deg g(x).$$

Thus $q_1(x) = q_2(x)$, whence also $r_1(x) = r_2(x)$. □

We shall find other rings which have a similar division algorithm, so it is worth introducing an abstract definition which encompasses them all. We do not have the notion of being positive in a general integral domain, so we shall have to replace $0 \leq r < |n|$ in the statement of the division algorithm for integers by $|r| < |n|$. We now lose the uniqueness of q, r because, for example, $13 = 3 \times 5 + (-2)$ so that, when dividing 13 by 5, we can obtain either $(q, r) = (2, 3)$ or $(3, -2)$. Thus we cannot insist on the uniqueness of the 'quotient' and 'remainder' in general.

What figures in the division algorithm on $\mathbb{Z}$ is the non-negative-integer-valued function $|\ |$ defined on $\mathbb{Z}$. Note that $|ab| = |a| |b|$ for all $a, b \in \mathbb{Z}$, so that, when $a, b \neq 0$, $|a| \leq |ab|$; or, put another way,

$$a|c \Rightarrow |a| \leq |c| \text{ for } a, c \neq 0.$$

For $k[x]$, this function is replaced by the degree function. There is the slight complication here that the zero polynomial has degree $-\infty$. Note that, for $f(x), g(x) \in k[x]$,

$$\deg(f(x)g(x)) = \deg f(x) + \deg g(x),$$

so that, when $f(x), g(x) \neq 0$,

$$\deg(f(x)) \leq \deg(f(x)g(x));$$

or, put another way,

$$f(x)|h(x) \Rightarrow \deg f(x) \leq \deg h(x) \text{ for } f(x), h(x) \neq 0.$$

The most convenient form of the definition is as follows:

2.3 Euclidean domains

Definition 2.3.3 The integral domain R is said to be a 'Euclidean domain' if there is a function $\partial \colon R\setminus\{0\} \to \{0, 1, 2, \ldots\}^\dagger$, called the 'degree function of R', such that
  (1) whenever a, b are non-zero elements of R such that $a|b$, then $\partial(a) \leqslant \partial(b)$,
  (2) given $a, b \in R$ with $b \neq 0$, there exist $q, r \in R$ such that
$$a = qb + r,$$
where either $r = 0$ or, if $r \neq 0$, $\partial(r) < \partial(b)$.

Of course, if $a \in R$, $a \neq 0$, we call the non-negative integer $\partial(a)$ the *degree* of a. We refer to property (2) of the definition as the *division algorithm* on the Euclidean domain R.

It should be clear that $\mathbb{Z}$ and $k[x]$, where k is any field, are Euclidean domains with $\partial(a) = |a|$ in $\mathbb{Z}$ and $\partial(f(x)) = \deg f(x)$ in $k[x]$. Also, any field k is an (uninteresting) example of a Euclidean domain with degree function defined by $\partial(a) = 0$ for all $a \in k$, $a \neq 0$. Note that, given $a, b \in k$ with $b \neq 0$, then $a = (ab^{-1})b$, so that we can always take $r = 0$ in the division algorithm on k.

For another example of a Euclidean domain, we turn to the ring $\mathbb{Z}[i]$ of Gaussian integers. We know from Section 1.4 that these form a subring of $\mathbb{C}$ and so form an integral domain using the addition and multiplication of complex numbers. But more:

Theorem 2.3.4 *The ring $\mathbb{Z}[i]$ of Gaussian integers is a Euclidean domain.*

Proof We define the 'norm function' N on $\mathbb{Z}[i]$ by
$$N(\alpha) = |\alpha|^2 \quad (\alpha \in \mathbb{Z}[i]).$$
Thus, for $\alpha \in \mathbb{Z}[i]$, $N(\alpha)$ is a non-negative integer which is equal to zero if and only if $\alpha = 0$. The degree function on $\mathbb{Z}[i]$ will just be the function N applied to $\mathbb{Z}[i]\setminus\{0\}$. Note that, for $\alpha, \beta \in \mathbb{Z}[i]$,
$$N(\alpha\beta) = |\alpha\beta|^2 = |\alpha|^2 |\beta|^2 = N(\alpha)N(\beta),$$

---

† $R\setminus\{0\}$ denotes the set of non-zero elements of R.

so that, when $\alpha, \beta \neq 0$, $N(\alpha) \leq N(\alpha\beta)$ and the first condition of Definition 2.3.3 follows from this.

Now consider $\alpha, \beta \in \mathbb{Z}[i]$ with $\beta \neq 0$. We can write
$$\alpha\beta^{-1} = \lambda + i\mu,$$
where $\lambda, \mu \in \mathbb{Q}$. Each rational number is within $\frac{1}{2}$ of an integer, so we can write
$$\lambda = \lambda_1 + \lambda_2, \mu = \mu_1 + \mu_2,$$
where $\lambda_1, \mu_1 \in \mathbb{Z}$ and $\lambda_2, \mu_2 \in \mathbb{Q}$ with $|\lambda_2| \leq \frac{1}{2}, |\mu_2| \leq \frac{1}{2}$. Now
$$\alpha\beta^{-1} = (\lambda_1 + i\mu_1) + (\lambda_2 + i\mu_2),$$
so that
$$\alpha = (\lambda_1 + i\mu_1)\beta + (\lambda_2 + i\mu_2)\beta.$$
Put $\lambda_1 + i\mu_1 = \gamma, (\lambda_2 + i\mu_2)\beta = \phi$, so that
$$\alpha = \gamma\beta + \phi.$$
Then $\phi = \alpha - \gamma\beta \in \mathbb{Z}[i]$ because $\alpha, \beta, \gamma \in \mathbb{Z}[i]$, and
$$N(\phi) = (\lambda_2^2 + \mu_2^2)|\beta|^2 \leq (\tfrac{1}{4} + \tfrac{1}{4})|\beta|^2 < |\beta|^2 = N(\beta).$$
The second condition of Definition 2.3.3 thus holds, and $\mathbb{Z}[i]$ is a Euclidean domain. □

The proof of Theorem 2.3.4 can be amended to show that $\mathbb{Z}[\sqrt{-2}]$ is a Euclidean domain (see Exercise 2.3.2); but a similar argument fails, for example, with $\mathbb{Z}[\sqrt{-3}]$. Not every ring of the form $\mathbb{Z}[\sqrt{n}]$, where $n \neq 0, 1$ and is $\pm$ (a product of distinct primes)[†], is a Euclidean domain. For example, we saw in Section 2.2 that $\mathbb{Z}[\sqrt{-5}]$ is not a UFD so, in consequence of what follows, it is not a Euclidean domain either. These rings arise in algebraic number theory as the *rings of algebraic integers* of certain number fields. In that context, it turns out that, when $n \equiv 1 \pmod 4$, we are not looking at the right ring; we should be looking at the ring $\mathbb{Z}[\frac{1}{2}(1 + \sqrt{n})]$. Note that
$$(\tfrac{1}{2}(1+\sqrt{n}))^2 = \tfrac{1}{4}(1 + 2\sqrt{n} + n) = \frac{1+\sqrt{n}}{2} + \frac{n-1}{4}$$

[†] i.e. 'square-free'.

2.3 Euclidean domains

and $\frac{1}{4}(n-1) \in \mathbb{Z}$. It follows from this that the elements of $\mathbb{Z}[\frac{1}{2}(1+\sqrt{n})]$ are all numbers (real if $n>0$ and complex if $n<0$) of the form $x+\frac{1}{2}y(1+\sqrt{n})$, where $x, y \in \mathbb{Z}$ (or, equivalently, all numbers of the form $a+b\sqrt{n}$, where a, b are either both integers or both halves of odd integers). Thus, for example, the ring $\mathbb{Z}[\sqrt{5}]$ in Exercise 2.2.2 is not the right ring from the point of view of number theory, because $5 \equiv 1 \pmod 4$. We do have the right ring when $n \equiv 2$ or $3 \pmod 4$, and n cannot be congruent to $0 \pmod 4$ because it is square-free.

The discussion of which of these rings of algebraic integers are Euclidean domains (using the degree function $\partial(a+b\sqrt{n}) = |a^2 - nb^2|$) is beyond the scope of this book. It turns out that, for negative values of n, when the rings consist of complex numbers, we obtain Euclidean domains for the values $n = -1, -2, -3, -7, -11$ and for these values alone. For positive values of n, when the rings consist of real numbers, we obtain Euclidean domains for the values $n = 2, 3, 5, 6, 7, 11, 13, 17, 19, 21, 29, 33, 37, 41, 57, 73$ and for these values alone. We refer readers to Hardy and Wright's *An Introduction to the Theory of Numbers*, fourth edition, Chapter 14.

To come back to earth, we need the following elementary result.

Theorem 2.3.5 Let R be a Euclidean domain with degree function ∂, and let a, b be non-zero elements of R such that $a|b$ and $\partial(a) = \partial(b)$. Then a, b are associates.

Proof We divide a by b using the division algorithm on R to give $q, r \in R$ such that

$$a = qb + r,$$

where either $r = 0$ or, if $r \neq 0$, $\partial(r) < \partial(b)$. Now $a|b$, so $b = ac$ for some $c \in R$. Thus $r = a(1 - qc)$ and $a|r$. Thus, if $r \neq 0$, $\partial(a) \leq \partial(r) < \partial(b)$, contrary to assumption. Hence, $r = 0$ and $b|a$. Since $a|b$, this means that a, b are associates. □

Before we can establish that Euclidean domains are UFDs, we need to discuss the existence of 'greatest common divisors' (or 'highest common factors') of elements in a Euclidean domain.

Exercises 2.3

1. In $\mathbb{Z}[i]$, determine a quotient and remainder when $11 + 7i$ is divided by $2 + 5i$. Are these unique?
2. Prove that $\mathbb{Z}[\sqrt{-2}]$ is a Euclidean domain.
3. Show that, in a Euclidean domain, non-zero elements which are associates have the same degree.
4. Show that, in a Euclidean domain R, $\partial(a) \geq \partial(1)$ for all non-zero elements a of R, and that $\partial(a) = \partial(1)$ if and only if a is a unit.
5. Let $\omega = e^{\pi i/3}$, and let R be the subring $\mathbb{Z}[\omega]$ of $\mathbb{C}$. Define the degree function ∂ on R by $\partial(a) = |a|^2$ ($a \in R \setminus \{0\}$).

   (a) Show that R is a Euclidean domain.

   (b) Determine the elements of R of degree 1.

   (c) Determine the units of R.

   (d) Show that $i\sqrt{3}$ is an irreducible element of R.

   (e) Show that $i\sqrt{3}|(m + n\omega)$ in R, where $m, n \in \mathbb{Z}$, if and only if $3|(m-n)$ in $\mathbb{Z}$.

   (f) Show that, if $a \in R$ is not divisible by $i\sqrt{3}$, then $a^3 \equiv \pm 1 \pmod{9}$ in R, i.e. $a^3 \mp 1 = 9b$ for some $b \in R$.

(Gauss used the Euclidean domain R and the results of this exercise to establish Fermat's Last Theorem for the case $n = 3$, i.e. he proved that there do not exist positive integers x, y, z such that $x^3 + y^3 = z^3$. See, for example, *100 Great Problems of Elementary Mathematics* by Heinrich Dörrie.

2.4

Greatest common divisors

Consider the integers 16, 24, 40. Their greatest common divisor is 8 in that 8|16, 8|24, 8|40 and 8 is the largest integer to divide them all. The word 'largest' has no meaning in a general integral domain, where there may be no ordering of the elements. But note that the factors common to 16, 24, 40 are ± 1, ± 2, ± 4, ± 8, and these are also factors of 8. It is this property that forms the definition of greatest common divisors in general. In this sense, -8 will also be a greatest common divisor of 16, 24, 40. Of course, 8 and -8 only differ by a unit in $\mathbb{Z}$.

2.4 Greatest common divisors

Definition 2.4.1 Let R be an integral domain and let $a_1, \ldots, a_n \in R, n \geq 1$. We say that d is a 'greatest common divisor' or 'highest common factor' (GCD or HCF for short) of $a_1, \ldots, a_n$ if (1) $d|a_1, \ldots, d|a_n$ (i.e. d is a 'common divisor' or 'common factor' of $a_1, \ldots, a_n$), and (2) whenever $c|a_1, \ldots, c|a_n$, where $c \in R$, then $c|d$.

Remarks (1) If d is a GCD of $a_1, \ldots, a_n \in R$, then so is ud for every unit u of R. Conversely, suppose that d_1, d_2 are GCDs of $a_1, \ldots, a_n$. Then, from the definition, $d_1|d_2$ and $d_2|d_1$, so that $d_1 \sim d_2$. Thus we may say that GCDs, if they exist, 'are unique up to associates', or that they 'differ only by a unit'.

(2) Let d be a GCD of $a_1, \ldots, a_n \in R$, with $a_1, \ldots, a_n$ not all zero. Then $d \neq 0$. We can write $a_1 = da'_1, \ldots, a_n = da'_n$ for some $a'_1, \ldots, a'_n \in R$. Now obviously $1|a'_1, \ldots, 1|a'_n$. Suppose that $c|a'_1, \ldots, c|a'_n$, where $c \in R$. Then $dc|a_1, \ldots, dc|a_n$, so that $dc|d$. Since $d \neq 0$, we can deduce that $c|1$. Thus $a'_1, \ldots, a'_n$ have GCD 1.

(3) Suppose that $a_1, \ldots, a_n \in R$ have GCD d, and let b be a non-zero element of R. Suppose also that $ba_1, \ldots, ba_n$ possess a GCD, c say. Then $d|a_i$ for every i, so $bd|ba_i$ for every i, so $bd|c$, say $c = bdu$ for some $u \in R$. Also $c|ba_i$ for every i, so $bdu|ba_i$ for every i, so $du|a_i$ for every i, so $du|d$, so $bdu|bd$, i.e. $c|bd$. Hence $c \sim bd$. Thus, provided we know that $a_1, \ldots, a_n$ possess a GCD and that $ba_1, \ldots, ba_n$ possess a GCD, then we can say that

$$\text{GCD}(ba_1, \ldots, ba_n) = b\,\text{GCD}(a_1, \ldots, a_n).$$

This fact will be used later.

Theorem 2.4.2 Every two elements a, b of a Euclidean domain R possess a GCD. Moreover, such a GCD can be expressed in the form $sa + tb$, where $s, t \in R$.

Remark In Exercise 2.4.2, we show that, if every two elements of R possess a GCD, then so does every finite non-empty set of elements. An integral domain R in which every two elements possess a GCD is sometimes called a *GCD domain*. If, in addition, a GCD of every two elements a, b can always be expressed in the form $sa + tb$, where $s, t \in R$ (and this also translates to the corresponding statement for every finite collection of elements), then R is called a *Bézout domain*

48 Factorization

after E. Bézout (1730–83). Thus a Euclidean domain is a Bézout domain.

We shall prove Theorem 2.4.2 by means of an algorithm which goes back to Euclid, appropriately called *Euclid's algorithm*. This provides a method of constructing a GCD of two elements a, b; it also constructs two elements $s, t \in R$ such that $sa + tb$ is a GCD of a, b.

Proof of Theorem 2.4.2 If a, b are both zero, then their GCD is zero (it fits the definition) and we can take s, t to be anything we like. Suppose now that they are not both zero. We may as well suppose that $b \neq 0$. If $b|a$, then it is easy to see that a, b have GCD b, and we can take $s = 0, t = 1$. If $b \nmid a$, we can use the division algorithm on R to say that there exist $q_1, r_1 \in R$ such that

$$a = q_1 b + r_1, r_1 \neq 0 \text{ and } \partial(r_1) < \partial(b).$$

We next divide b by r_1. If $r_1 | b$, the process stops. Otherwise, there exist $q_2, r_2 \in R$ such that

$$b = q_2 r_1 + r_2, r_2 \neq 0 \text{ and } \partial(r_2) < \partial(r_1).$$

We next divide r_1 by r_2. If $r_2 | r_1$, the process stops. Otherwise, there exist $q_3, r_3 \in R$ such that

$$r_1 = q_3 r_2 + r_3, r_3 \neq 0 \text{ and } \partial(r_3) < \partial(r_2).$$

We continue in this way. The degrees of the successive remainder terms are non-negative integers and are strictly decreasing, so this procedure must stop at some stage. Let r_n be the last non-zero remainder, so that the last two divisions are

$$r_{n-2} = q_n r_{n-1} + r_n,$$

$$r_{n-1} = q_{n+1} r_n.$$

From the last division, we see that $r_n | r_{n-1}$. From the penultimate division, we now see that $r_n | r_{n-2}$ (because r_n divides both r_n and r_{n-1}). Working our way back through the divisions, we eventually see that $r_n | b, r_n | a$, so that r_n is a common factor of a, b.

Suppose now that $c|a, c|b$, where $c \in R$. From the first division $r_1 = a - q_1 b$, so that $c | r_1$. Then, from the second division, $c|b, c|r_1$ so that $c|(b - q_2 r_1)$, i.e. $c|r_2$. Continuing through the divisions, we eventually see that $c|r_n$. This shows that r_n is a GCD of a, b.

2.4 Greatest common divisors

Finally,

$$r_1 = a - q_1 b,$$
$$r_2 = b - q_2 r_1 = b - q_2(a - q_1 b) = -q_2 a + (1 + q_2 q_1)b,$$
$$r_3 = r_1 - q_3 r_2 = (a - q_1 b) - q_3[-q_2 a + (1 + q_2 q_1)b]$$
$$= (1 + q_3 q_2)a + (-q_1 - q_3 - q_3 q_2 q_1)b.$$

Continuing through the divisions, we eventually express r_n in the form $sa + tb$ for some $s, t \in R$. □

Example 2.4.3 Express the GCD of the integers 132, 630 in the form $132s + 630t$ where $s, t \in \mathbb{Z}$.

Solution We use Euclid's algorithm. We set out the successive divisions as follows:

```
        4
  132)630
      528    1
      102)132
          102    3
            30)102
                90    2
                12)30
                   24    2
                    6)12
                     12
                      0
```

The last non-zero remainder is 6, so the GCD of 132, 630 is 6. Working back through the divisions, we have that

$$6 = 30 - 2 \times 12$$
$$= 30 - 2(102 - 3 \times 30)$$
$$= 7 \times 30 - 2 \times 102$$
$$= 7(132 - 102) - 2 \times 102$$
$$= 7 \times 132 - 9 \times 102$$
$$= 7 \times 132 - 9(630 - 4 \times 132)$$
$$= 43 \times 132 - 9 \times 630,$$

50   Factorization

so we can take $s=43$, $t=-9$. (We could also take $s=43+630$, $t=-9-132$, for example, so that s, t are not unique.) □

We can also apply Euclid's algorithm in less familiar situations.

Example 2.4.4   Find the GCD of the Gaussian integers $7+3i$, $5-8i$.

Solution   We use the technique of the proof of Theorem 2.3.4, where we showed that the Gaussian integers form a Euclidean domain. We first divide one Gaussian integer by the other in $\mathbb{C}$, to give

$$\frac{5-8i}{7+3i} = \frac{(5-8i)(7-3i)}{(7+3i)(7-3i)} = \frac{11}{58} - \frac{71}{58}i.$$

The Gaussian integer 'nearest' to this is $-i$, so we take the quotient to be $-i$. This gives the first division in $\mathbb{Z}[i]$ to be

$$5-8i = -i(7+3i)+(2-i),$$

with $2-i$ as remainder. Next we divide $7+3i$ by $2-i$ in $\mathbb{C}$ to give

$$\frac{7+3i}{2-i} = \frac{(7+3i)(2+i)}{(2-i)(2+i)} = \frac{11}{5} + \frac{13}{5}i.$$

The Gaussian integer 'nearest' to this is $2+3i$, which we take as the quotient when $7+3i$ is divided by $2-i$ in $\mathbb{Z}[i]$, to give

$$7+3i = (2+3i)(2-i)-i.$$

This is the second division in $\mathbb{Z}[i]$. It is not necessary to carry out a further division, because the remainder in the second division is $-i$, a unit, and so will divide into $2-i$ with zero remainder. Hence the GCD of $7+3i$, $5-8i$ is $-i$.

Readers may like to carry out the same procedure but reversing the roles of $7+3i$, $5-8i$, so that the first step is to divide $7+3i$ by $5-8i$. The last non-zero remainder should come out to be -1, so that the GCD is -1. But $-i, -1$ are both units in $\mathbb{Z}[i]$, with respective inverses $i, -1$. Since GCDs are only defined up to associates, we may as well say that the GCD of $7+3i$, $5-8i$ in $\mathbb{Z}[i]$ is 1. □

2.4 Greatest common divisors

Definition 2.4.5 If two elements of an integral domain have GCD 1, then we say that they are 'coprime' (or 'relatively prime').

Thus the Gaussian integers $7+3i$, $5-8i$ are coprime.

We note that, if a,b are coprime elements of a Euclidean domain (or, more generally, of a Bézout domain) R, then there exist $s, t \in R$ such that

$$sa + tb = 1.$$

(It is an easy matter to see that the converse is also true – see Exercise 2.4.6.)

We shall need the following results in Section 2.8. The first will probably be familiar to readers, at least when $R = \mathbb{Z}$.

Theorem 2.4.6 Let R be a Euclidean domain and let $a, b, c \in R$. If $a|bc$ and a, b are coprime, then $a|c$.

Proof Suppose that $a|bc$ and that a, b are coprime. Then there exist $s, t \in R$ such that $sa + tb = 1$. Now $a|ac$ and $a|bc$, so $a|(sac + tbc)$, i.e. $a|c$. □

To illustrate the second result, consider the integers 24, 30, −36. We can factorize these into irreducibles in the UFD $\mathbb{Z}$ to give

$$24 = 2^3 \times 3, \quad 30 = 2 \times 3 \times 5, \quad -36 = 2 \times (-2) \times 3^2.$$

Then the GCD of these integers is 6 (or −6), because 6 divides the integers, and consideration of unique factorization shows that any integer dividing 24, 30, −36 must be of the form $\pm 2^\lambda \times 3^\mu$, where $\lambda = 0$ or 1, $\mu = 0$ or 1. In the proof which follows, we would begin by writing

$$24 = 1 \times 2^3 \times 3^1 \times 5^0$$
$$30 = 1 \times 2^1 \times 3^1 \times 5^1$$
$$-36 = (-1) \times 2^2 \times 3^2 \times 5^0.$$

Note that 1, −1 are units.

Theorem 2.4.7 In a UFD, every finite non-empty set of elements possesses a GCD.

Thus a UFD is a GCD domain.

Proof Let R be a UFD, and consider $a_1, \ldots, a_r \in R, r \geq 1$. In the manner illustrated above for $\mathbb{Z}$, we can factorize $a_1, \ldots, a_r$ into irreducibles as follows:

$$a_1 = u_1 p_1^{\lambda_{11}} p_2^{\lambda_{12}} \ldots p_n^{\lambda_{1n}},$$
$$a_2 = u_2 p_1^{\lambda_{21}} p_2^{\lambda_{22}} \ldots p_n^{\lambda_{2n}},$$
$$\ldots\ldots\ldots\ldots\ldots\ldots\ldots\ldots$$
$$a_r = u_r p_1^{\lambda_{r1}} p_2^{\lambda_{r2}} \ldots p_n^{\lambda_{rn}},$$

where $u_1, \ldots, u_r$ are units, $p_1, \ldots, p_n$ are non-associate irreducible elements of R and the λ_{ij} are non-negative integers. For $1 \leq j \leq n$, put

$$\lambda_j = \min\{\lambda_{1j}, \lambda_{2j}, \ldots, \lambda_{rj}\},$$

and write $d = p_1^{\lambda_1} p_2^{\lambda_2} \ldots p_n^{\lambda_n}$. Then $d | a_1, \ldots, d | a_r$. Suppose that $c \in R$ is such that $c | a_i$ for $1 \leq i \leq r$, say $a_i = b_i c$ for some $b_i \in R$. From unique factorization, we see that c must be of the form $c = u p_1^{\mu_1} p_2^{\mu_2} \ldots p_n^{\mu_n}$, where u is a unit and $\mu_1 \leq \lambda_{11}, \ldots, \lambda_{r1}, \ldots, \mu_n \leq \lambda_{1n}, \ldots, \lambda_{rn}$, i.e. $\mu_1 \leq \lambda_1, \ldots, \mu_n \leq \lambda_n$. Thus $c | d$ and d is a GCD of $a_1, \ldots, a_r$. □

Exercises 2.4
1. Show that 3 and $2 + \sqrt{-5}$ have a GCD in $\mathbb{Z}[\sqrt{-5}]$, but that it is not possible to express this GCD in the form $3\alpha + (2 + \sqrt{-5})\beta$, where $\alpha, \beta \in \mathbb{Z}[\sqrt{-5}]$.
2. Show that, if every two elements of R possess a GCD, then so does every finite non-empty set of elements. Show also that, if every two elements a, b of R have a GCD which is expressible in the form $sa + tb$, where $s, t \in R$, then every finite set $a_1, \ldots, a_n$ ($n \geq 1$) of elements of R have a GCD which is expressible in the form $s_1 a_1 + \cdots + s_n a_n$, where $s_1, \ldots, s_n \in R$.
3. Find the GCD of $-5 + 2\sqrt{-2}, 1 + 5\sqrt{-2}$ in $\mathbb{Z}[\sqrt{-2}]$, and express it in the form

$$(-5 + 2\sqrt{-2})\alpha + (1 + 5\sqrt{-2})\beta,$$

where $\alpha, \beta \in \mathbb{Z}[\sqrt{-2}]$.
4. Find the monic GCD of

$$f(x) = x^5 + \bar{4}x, \quad g(x) = x^6 + x^5 + \bar{3}x^4 + x^3 + x^2 + x + \bar{2}$$

2.5 Prime elements

in $\mathbb{Z}_5[x]$ (monic = leading coefficient $\bar{1}$), and express it in the form $s(x)f(x) + t(x)g(x)$, where $s(x), t(x) \in \mathbb{Z}_5[x]$. (Note that, in the expression for $f(x)$, $\bar{1}x^5$ has been written simply as x^5; this is done consistently. Also, terms with zero coefficients are omitted.)

5. Find a solution of the equation
$$(4 + 9i)\alpha + (2 + 7i)\beta = 3 + i$$
in Gaussian integers α, β.

6. Let a, b be elements of an arbitrary integral domain R for which there exist $s, t \in R$ such that $sa + tb = 1$. Show that a, b are coprime.

7. Let R be a Euclidean domain and let $a, b, c \in R$. Show that, if $a|c$, $b|c$ and a, b are coprime, then $ab|c$.

2.5

Prime elements

Readers may know that, if p is a prime number in $\mathbb{Z}$, then, whenever $p|ab$, where $a, b \in \mathbb{Z}$, either $p|a$ or $p|b$. We use this property to define a prime element in an arbitrary integral domain.

Definition 2.5.1 Let R be an integral domain and let $p \in R$. We say that p is 'prime' if (1) $p \neq 0$, (2) p is not a unit, and (3) whenever $p|ab$, where $a, b \in R$, then either $p|a$ or $p|b$.

In $\mathbb{Z}$, the prime elements are, in fact, the elements of the form $\pm p$, where p is a prime number. These are the same as the irreducible elements of $\mathbb{Z}$. In the next theorem, we shall show that prime and irreducible elements are one and the same in a UFD (such as $\mathbb{Z}$).

To see that, in general, the prime and the irreducible elements are not the same, consider $3 \in \mathbb{Z}[\sqrt{-5}]$. We know from Section 2.2 that 3 is irreducible. Also,
$$3 \times 3 = (2 + \sqrt{-5})(2 - \sqrt{-5}),$$
so that $3|(2 + \sqrt{-5})(2 - \sqrt{-5})$. But $3 \nmid (2 + \sqrt{-5})$ in $\mathbb{Z}[\sqrt{-5}]$, since otherwise $2 + \sqrt{-5} = 3(a + b\sqrt{-5})$ for some $a, b \in \mathbb{Z}$, whence $2 = 3a$,

which is impossible. Also $3 \nmid (2-\sqrt{-5})$. Thus 3 is not prime in $\mathbb{Z}[\sqrt{-5}]$. Of course, $\mathbb{Z}[\sqrt{-5}]$ is not a UFD.

Notes (1) An easy extension of the definition of a prime element of the integral domain R gives that, if p is prime and $p|a_1 a_2 \ldots a_s$, where $a_1, \ldots, a_s \in R$ and $s \geqslant 1$, then $p|a_i$ for some i.

(2) If p is a prime element of an integral domain R and u is a unit, then it is easy to see that up is also a prime element of R.

Theorem 2.5.2 Let R be an integral domain. Then (1) every prime element of R is irreducible, (2) if R is a Euclidean domain, every irreducible element is prime, (3) if R is a UFD, every irreducible element is prime.

Remark The result for which we are aiming is that a Euclidean domain is a UFD (Theorem 2.6.1). When once we have that result, statement (2) of Theorem 2.5.2 is included in statement (3). However, (2) is an important step in proving Theorem 2.6.1, so we cannot dispense with it at this stage. We shall need (3) to prove Theorem 2.8.9.

Proof of Theorem 2.5.2 (1) Let p be a prime element of R, and consider the factorization $p = ab$, where $a, b \in R$. Then $p|ab$, so either $p|a$ or $p|b$. But also $a|p$, $b|p$, so either $a \sim p$ or $b \sim p$. Thus p has no proper factors and so is irreducible.

(2) Now suppose that R is a Euclidean domain, and let p be an irreducible element of R. Suppose that $a, b \in R$ are such that $p|ab$ but $p \nmid a$. By Theorem 2.4.2, a, p have a GCD. Now only units and associates of p divide p because p is irreducible, and associates of p do not divide a. Hence a, p are coprime. It follows from Theorem 2.4.6 that $p|b$. Thus p is prime.

(3) Now suppose that R is a UFD, and let p be an irreducible element of R. Suppose that $a, b \in R$ are such that $p|ab$. Then there exists $c \in R$ such that $ab = pc$. If we now factorize a, b, c into irreducibles, we see by uniqueness that there must be an irreducible factor of a or b which is an associate of p, so that either $p|a$ or $p|b$. Hence p is prime. □

2.6 Euclidean domains are UFDs

Remark We have in fact proved in (2) that, in a Bézout domain, every irreducible element is prime.

Exercises 2.5

1. Find an irreducible element in $\mathbb{Z}[\sqrt{5}]$ which is not prime.
2. Let p be a positive integer and let $k \in \mathbb{Z}$ with $0 \leq k \leq p$. Show that the binomial coefficient

$$\binom{p}{k} = \frac{p!}{k!(p-k)!}$$

is an integer, and that, when p is prime and $1 \leq k \leq p-1$, it is divisible by p.

2.6

Euclidean domains are UFDs

We are now in a position to prove the result for which we have been aiming.

Theorem 2.6.1 Every Euclidean domain is a UFD.

Proof Let R be a Euclidean domain with degree function ∂. We first show that every non-zero, non-unit element of R can be factorized into irreducibles. Suppose the contrary. Then there exists a non-zero, non-unit element with no factorization into irreducibles and, by the well-ordering of the non-negative integers, we can choose such an element of smallest degree, say a. Now a cannot be irreducible, for otherwise it can certainly be factorized into irreducibles. Hence, there exist b, c in R, not units and not associates of a, such that $a = bc$. Since $b | a$, $\partial(b) \leq \partial(a)$. But, since b and a are not associates, Theorem 2.3.5 tells us that $\partial(b) \neq \partial(a)$. Hence, $\partial(b) < \partial(a)$. Similarly, $\partial(c) < \partial(a)$. It follows by the minimality of $\partial(a)$ that b, c can both be factorized as products of finitely many irreducibles. If we put these two factorizations together, we obtain a factorization of a into irreducibles. This gives the required contradiction.

To establish uniqueness, suppose that

$$p_1 p_2 \cdots p_s = q_1 q_2 \cdots q_t,$$

where the p_i, q_j are irreducibles. By Theorem 2.5.2(2), p_s is prime, and $p_s|(q_1q_2\ldots q_t)$, so that $p_s|q_j$ for some j. Since the order of the factors is unimportant, we can reorder the q_j's as necessary and so suppose that $p_s|q_t$. But q_t is irreducible, and so has no proper factors. Thus $p_s \sim q_t$, say $q_t = up_s$, where u is some unit. We can now cancel p_s to give

$$p_1 p_2 \ldots p_{s-1} = (uq_1)q_2 \ldots q_{t-1}.$$

Note that uq_1 is irreducible. We now repeat this procedure with p_{s-1} in place of p_s, then p_{s-2}, and so on. We shall run out of the p_i's at the same time as we run out of irreducibles on the right-hand side, otherwise we would end up with a product of irreducibles equal to 1, which is impossible. Hence $s = t$ and we can rearrange the q_j's such that $p_i \sim q_i$ for $1 \leq i \leq s$. This proves the uniqueness. □

It follows from Theorem 2.6.1 that $\mathbb{Z}$, the polynomial rings $k[x]$, where k is any field, and $\mathbb{Z}[i]$ are all UFDs.

2.7

The two-squares theorem

An interesting application of factorization in the ring $\mathbb{Z}[i]$ of Gaussian integers is to prove the famous 'two-squares theorem' in number theory, due originally to Fermat. It may be asked which prime numbers (we consider non-primes later) can be expressed as sums of two squares†. Thus, $2 = 1^2 + 1^2$, $5 = 2^2 + 1^2$, $13 = 3^2 + 2^2$, $17 = 4^2 + 1^2$, $29 = 5^2 + 2^2$, but it is impossible to express 3, 7, 11, 19, 23 in this way. Fermat's result tells us exactly which primes are the sums of two squares.

We first determine the units of the ring $\mathbb{Z}[i]$.

Theorem 2.7.1 Let $\alpha \in \mathbb{Z}[i]$. Then the following statements are equivalent:
  (1) α is a unit of $\mathbb{Z}[i]$;

---

† We use the phrase 'sums of two squares' always to mean an expression of the form $a^2 + b^2$, where $a, b \in \mathbb{Z}$; we do not allow $3 = 2^2 + i^2$, for example.

2.7 The two-squares theorem

(2) $N(\alpha)=1$ (where $N(a+ib)=a^2+b^2$ with $a,b\in\mathbb{Z}$);
(3) $\alpha=\pm 1$ or $\pm i$.

Proof Assume (1). Then $\alpha\alpha^{-1}=1$, so $N(\alpha)N(\alpha^{-1})=N(\alpha\alpha^{-1})=N(1)=1$, whence $N(\alpha)=1$.

Now assume (2), and write $\alpha=a+ib$, where $a,b\in\mathbb{Z}$. Then $a^2+b^2=1$, so either $a=\pm 1, b=0$ or $a=0, b=\pm 1$. Thus either $\alpha=\pm 1$ or $\pm i$.

It is easy to see that $\pm 1, \pm i$ are units, so (3)$\Rightarrow$(1). □

The next result gives Fermat's result that every prime number which is congruent to 1 (mod 4) is a sum of two squares.

Theorem 2.7.2 Let p be a prime number in $\mathbb{Z}$. Then the following statements are equivalent:
(1) $p\equiv 3$ (mod 4);
(2) p is irreducible as a Gaussian integer;
(3) p is not a sum of two squares.

Proof First assume that (1) is false. If $p=2$, then $p=(1+i)(1-i)$, and neither $1+i$ nor $1-i$ is a unit, so p is not irreducible. Now suppose that $p\equiv 1$ (mod 4). By Wilson's Theorem (Theorem 1.7.4),

$$1\times 2\times\ldots\times\left(\frac{p-1}{2}\right)\times\left(\frac{p+1}{2}\right)\times\left(\frac{p+3}{2}\right)$$
$$\times\ldots\times(p-1)\equiv -1\ (\text{mod } p).$$

If we subtract p from the later terms in the product, then the congruence is preserved and

$$1\times 2\times\ldots\times\left(\frac{p-1}{2}\right)\times\left(-\frac{p-1}{2}\right)\times\left(-\frac{p-3}{2}\right)$$
$$\times\ldots\times(-1)\equiv -1\ (\text{mod } p),$$

whence

$$(-1)^{(p-1)/2}\left(1\times 2\times\ldots\times\left(\frac{p-1}{2}\right)\right)^2\equiv -1\ (\text{mod } p).$$

Put $q = ((p-1)/2)!$. Then, since $(p-1)/2$ is even because $p \equiv 1 \pmod{4}$,
$$q^2 \equiv -1 \pmod{p}.$$
Thus $p|(q^2+1)$ in $\mathbb{Z}$ and so also in the ring $\mathbb{Z}[i]$. Thus we have $p|(q+i)(q-i)$ in $\mathbb{Z}[i]$. Suppose $p|(q+i)$ in $\mathbb{Z}[i]$. Then there exist $y, z \in \mathbb{Z}$ such that $q+i = p(y+iz)$, from which $1 = pz$, which is impossible. Thus $p \nmid (q+i)$ in $\mathbb{Z}[i]$, and similarly $p \nmid (q-i)$ in $\mathbb{Z}[i]$. This means that p is not a prime element of $\mathbb{Z}[i]$. But $\mathbb{Z}[i]$ is a Euclidean domain, so, by Theorem 2.5.2(2), p is not irreducible in $\mathbb{Z}[i]$. Thus, if (1) is false, then (2) is false.

Now assume that (2) is false. Certainly p is not a unit in $\mathbb{Z}[i]$, so we can write
$$p = (a+ib)(c+id),$$
where $a, b, c, d \in \mathbb{Z}$ and $a+ib, c+id$ are not units of $\mathbb{Z}[i]$. Taking norms, we have
$$p^2 = (a^2+b^2)(c^2+d^2).$$
Now $a^2+b^2, c^2+d^2 > 1$ because $a+ib, c+id$ are not units of $\mathbb{Z}[i]$, so the only possibility is that $p = a^2+b^2 \,(= c^2+d^2)$ and (3) is false.

Finally, suppose that (3) is false. Then $p = a^2+b^2$ for some $a, b \in \mathbb{Z}$. Now $a, b \equiv 0$ or $1 \pmod 2$, so $a^2 \equiv 0$ or $1 \pmod 4$. Thus $p \equiv 0, 1$ or $2 \pmod 4$ (0 is actually impossible because $4 \nmid p$), so $p \not\equiv 3 \pmod 4$ and (1) is false. □

The next result describes all the irreducible Gaussian integers.

Theorem 2.7.3 The irreducible Gaussian integers are precisely
  (1) $\pm p, \pm ip$, where p is a prime number in $\mathbb{Z}$ and $p \equiv 3 \pmod 4$,
  (2) elements of prime norm[†].

Proof That Gaussian integers of the form (1) are irreducible follows from Theorem 2.7.2 (and 2.7.1). Let $\pi \in \mathbb{Z}[i]$ with $N(\pi)$ prime. Then π is not a unit (nor zero). Write $\pi = \alpha\beta$, where $\alpha, \beta \in \mathbb{Z}[i]$. Then $N(\pi) = N(\alpha)N(\beta)$, so either $N(\alpha) = 1$ or $N(\beta) = 1$. Thus α or β is a unit, so π is irreducible.

---

† This is the usual norm given by $N(a+ib) = a^2+b^2$, where $a, b \in \mathbb{Z}$.

2.7 The two-squares theorem

Conversely, suppose that π is an irreducible Gaussian integer, and write $\pi = a + ib$, where $a, b \in \mathbb{Z}$. Suppose that $b = 0$, so that $\pi = a$. If a is not $\pm$ a prime number, then a can be properly factorized in $\mathbb{Z}$, and such a factorization will also be a proper factorization in $\mathbb{Z}[i]$. Hence, $\pi = \pm p$, where p is a prime number in $\mathbb{Z}$. Now p must be irreducible as a Gaussian integer so, by Theorem 2.7.2, $p \equiv 3 \pmod 4$. Suppose next that $a = 0$, so that $\pi = ib$. By a similar argument, we see that $\pi = \pm ip$, where p is a prime number in $\mathbb{Z}$ and $p \equiv 3 \pmod 4$.

Now suppose that $\pi = a + ib$ is irreducible and $a, b \neq 0$. We claim that $N(\pi)$ is prime. Suppose not. Then $N(\pi) = a^2 + b^2 = cd$ for some $c, d \in \mathbb{Z}$, $c, d > 1$. Then

$$a^2 + b^2 = (a + ib)(a - ib) = cd.$$

Now $a + ib$ is irreducible in $\mathbb{Z}[i]$. Moreover, since any factorization of $a - ib$ in $\mathbb{Z}[i]$ will give a corresponding factorization of $a + ib$ by conjugating all the factors, $a - ib$ is also irreducible in $\mathbb{Z}[i]$. If c, d were irreducible in $\mathbb{Z}[i]$, we would have two essentially different factorizations of $a^2 + b^2$ into irreducibles in $\mathbb{Z}[i]$, contradicting the fact that $\mathbb{Z}[i]$ is a UFD. If one of c, d is not irreducible in $\mathbb{Z}[i]$ (neither is a unit), we shall have a product of two irreducibles equal to a product of more than two irreducibles, which again contradicts uniqueness of factorization. Hence, $N(\pi)$ is prime. □

The next result shows that a prime number in $\mathbb{Z}$ not congruent to 3 (mod 4) can be expressed in only one way as a sum of two squares, except for the order of the terms in the sum.

Theorem 2.7.4 Let p be a prime number in $\mathbb{Z}$ such that

$$p = a^2 + b^2 = c^2 + d^2,$$

where a, b, c, d are positive integers (so that $p \not\equiv 3 \pmod 4$). Then either $a = c, b = d$ or $a = d, b = c$.

Proof We have

$$(a + ib)(a - ib) = (c + id)(c - id)$$

and $N(a \pm ib) = N(c \pm id) = p$, a prime, so that, by Theorem 2.7.3, $a \pm ib, c \pm id$ are irreducible elements of $\mathbb{Z}[i]$. It follows from unique factorization that $a+ib \sim c \pm id$, i.e. $a+ib$ is one of the eight elements $\pm(c \pm id)$, $\pm i(c \pm id)$. But $a, b, c, d > 0$, so $a+ib = c+id$ or $d+ic$, i.e. either $a=c$, $b=d$ or $a=d$, $b=c$. $\square$

We next consider which integers, not just prime numbers, are sums of two squares. First an elementary observation.

Theorem 2.7.5 The product of two integers, each of which is the sum of two squares, is itself the sum of two squares.

Proof Let $a, b, c, d \in \mathbb{Z}$. Then

$$\begin{aligned}(a^2+b^2)(c^2+d^2) &= (a+ib)(a-ib)(c+id)(c-id)\\ &= (a+ib)(c+id)(a-ib)(c-id)\\ &= (ac-bd)^2 + (ad+bc)^2. \quad \square\end{aligned}$$

Theorem 2.7.6 Let $n \in \mathbb{Z}$, $n > 1$. Then n is a sum of two squares if and only if its prime factors congruent to 3 (mod 4) occur to even powers in its prime factorization.

Thus 18 is a sum of two squares ($18 = 3^2 + 3^2$) because $18 = 2 \times 3^2$, but 14 is not because $14 = 2 \times 7$ and its prime factor 7, which is congruent to 3 (mod 4), occurs to an odd power (namely 1) in the prime factorization of 14.

Proof of Theorem 2.7.6 Suppose that the prime factors of n which are congruent to 3 (mod 4) occur to even powers in its prime factorization. Then we can write

$$n = p_1^2 p_2^2 \ldots p_r^2 p_{r+1} p_{r+2} \ldots p_s,$$

where $p_1, \ldots, p_r$ are prime numbers congruent to 3 (mod 4) and $p_{r+1}, \ldots, p_s$ are prime numbers not congruent to 3 (mod 4). Then $p_1^2, \ldots, p_r^2, p_{r+1}, \ldots, p_s$ are all sums of two squares ($p_j^2 = p_j^2 + 0^2$ and, for $r+1 \leq l \leq s$, p_l is a sum of two squares by Theorem 2.7.2). Thus, by an obvious extension of Theorem 2.7.5, n is a sum of two squares.

Conversely, suppose that n is a sum of two squares, say $n = a^2 + b^2$, where $a, b \in \mathbb{Z}$. Then

$$n = (a+ib)(a-ib).$$

2.7 The two-squares theorem

Since $\mathbb{Z}[i]$ is a UFD, we can factorize $a \pm ib$ into irreducibles; using Theorem 2.7.3, we write

$$a+ib = \pi_1 \pi_2 \ldots \pi_r(a_1+ib_1)\ldots(a_s+ib_s),$$

where, for $1 \leq j \leq r$, $\pi_j = \pm p_j$ or $\pm ip_j$, p_j being a prime number in $\mathbb{Z}$ congruent to 3 (mod 4), and $N(a_l+ib_l) = a_l^2 + b_l^2$ is prime for $1 \leq l \leq s$. If we take conjugates, we have

$$a-ib = \bar{\pi}_1 \bar{\pi}_2 \ldots \bar{\pi}_r(a_1-ib_1)\ldots(a_s-ib_s).$$

Thus

$$n = |\pi_1|^2 |\pi_2|^2 \ldots |\pi_r|^2 (a_1^2+b_1^2)\ldots(a_s^2+b_s^2).$$

Now, for $1 \leq j \leq r$, $|\pi_j| = p_j$ and so is a prime number in $\mathbb{Z}$ congruent to 3 (mod 4). Further, by Theorem 2.7.2, the prime numbers $a_l^2 + b_l^2$ ($1 \leq l \leq s$) are not congruent to 3 (mod 4). Thus every prime factor of n congruent to 3 (mod 4) occurs to an even power in its prime factorization. □

Exercises 2.7

1. Express 3965 as a sum of two squares in as many ways as you can. (Calculators are not allowed!)
2. My internal telephone number is 4430 and a colleague's is 4437. Factorize these into irreducibles in $\mathbb{Z}[i]$ and, if possible, express them as the sum of two squares.
3. Factorize the Gaussian integers $3+i$, $4+3i$, $75+28i$ into irreducibles.
4. For which prime numbers p does the congruence $x^2 \equiv -1 \pmod{p}$ have a solution in $\mathbb{Z}$?
5. Let p be a prime number. Show that the set F of all matrices of the form

$$\begin{bmatrix} \bar{a} & \bar{b} \\ -\bar{b} & \bar{a} \end{bmatrix},$$

where $a, b \in \mathbb{Z}$ and $\bar{a}, \bar{b} \in \mathbb{Z}_p$, the field of residue-classes of integers modulo p, is a commutative ring under matrix addition and multiplication, and that F is a field if and only if $p \equiv 3 \pmod{4}$. How many elements does F possess?

62   *Factorization*

An extra problem: the problem of the Yellow Brick Road This appeared as the 1976 Christmas problem in the *Yorkshire Post*, and is reproduced here by courtesy of the *Yorkshire Post*.

'Pooh', said the wizard, 'is where we are now. The railway starts here and runs in a straight line via 39 intermediate and equally spaced stations to the terminus at Oz.

'Unfortunately, the railwaymen are on strike, so you'll have to go by bus. The bus goes along the Yellow Brick Road, which runs in a straight line from here to the outlying village of Bah, where it turns through a right-angle and goes in a straight line back to the first railway station after Pooh. From there it goes in a straight line to the next outlying village, where it turns through a right-angle and proceeds in a similar zig-zag fashion all the way to Oz, alternately calling at railway stations and outlying villages. Each of the 80 straight stretches of road is a different whole number of miles long. Rail distances are also whole numbers of miles.

'The fare is one ozzle per mile, but you needn't be alarmed, as all distances are as short as they can be.'

I was alarmed, and it turned out that I had good reason to be. My money was running short, for the Wizardry of Oz had been suffering from hyper-inflation lately.

Unfortunately the Wizard had vanished before I could ask him the vital question, HOW LONG IS THE YELLOW BRICK ROAD?

2.8

Factorization of polynomials

In this section we shall investigate factorization specifically for polynomials. After the integers, with the Fundamental Theorem of Arithmetic, this is the second situation in which factorization arose historically.

We shall first recall some elementary results about roots of polynomials. Initially, these are valid when the coefficients belong to an arbitrary commutative ring R. Let $\alpha \in R$, and consider the polynomial $f(x) \in R[x]$, say

$$f(x) = a_0 + a_1 x + \ldots + a_n x^n$$

2.8 Factorization of polynomials

with $a_0, a_1, \ldots, a_n \in R$. Now

$$\begin{aligned} f(x) - f(\alpha) &= a_1(x-\alpha) + a_2(x^2-\alpha^2) + \ldots \\ &\quad + a_n(x^n - \alpha^n) \\ &= a_1(x-\alpha) + a_2(x-\alpha)(x+\alpha) + \ldots \\ &\quad + a_n(x-\alpha)(x^{n-1} + \alpha x^{n-2} + \ldots + \alpha^{n-1}) \\ &= (x-\alpha)g(x) \end{aligned}$$

for some polynomial $g(x) \in R[x]$. Thus

$$f(x) = (x-\alpha)g(x) + f(\alpha).$$

In fact, when R is a field, $g(x)$ is the quotient and $f(\alpha)$ is the remainder when $f(x)$ is divided by $x - \alpha$. This result is referred to as the *Remainder Theorem*, and could easily have been obtained from the division algorithm for polynomials (when R is a field).

Returning to the situation when R is an arbitrary commutative ring, we see that, if $f(\alpha) = 0$, then $x - \alpha$ is a factor of $f(x)$. Conversely, suppose that $x - \alpha$ is a factor of $f(x)$, say $f(x) = (x-\alpha)h(x)$ for some $h(x) \in R[x]$. Then

$$(x-\alpha)(h(x) - g(x)) = f(\alpha).$$

If $h(x) - g(x) \neq 0$, then the left-hand side will have positive degree, and so cannot be equal to $f(\alpha)$. Hence, $h(x) - g(x) = 0$, and so also $f(\alpha) = 0$.

Definition 2.8.1 Let R be a commutative ring, let $f(x)$ be a polynomial in $R[x]$ and let $\alpha \in R$. We say that α is a 'root' of $f(x)$ if $f(\alpha) = 0$.

We have thus proved the following result:

Theorem 2.8.2 (**The Factor Theorem**) Let R be a commutative ring, let $f(x)$ be a polynomial in $R[x]$ and let $\alpha \in R$. Then the following statements are equivalent:
  (1) $x - \alpha$ is a factor of $f(x)$;
  (2) α is a root of $f(x)$. □

We can immediately extend Theorem 2.8.2 when R is an integral domain.

Corollary 2.8.3 Let R be an integral domain, let $\alpha_1, \ldots, \alpha_r (r \geqslant 1)$ be distinct elements of R, and let $f(x) \in R[x]$. Then the following statements are equivalent:
  (1) $(x-\alpha_1)(x-\alpha_2)\ldots(x-\alpha_r)$ is a factor of $f(x)$;
  (2) $\alpha_1, \alpha_2, \ldots, \alpha_r$ are roots of $f(x)$.

Proof If (1) holds then, for each i, $x - \alpha_i$ is a factor of $f(x)$ so, by Theorem 2.8.2, α_i is a root of $f(x)$. Thus (1)$\Rightarrow$(2).

We prove that (2)$\Rightarrow$(1) by induction on r. The case $r=1$ is just (2)$\Rightarrow$(1) in Theorem 2.8.2. Now let $r > 1$, assume the result for $r-1$ and suppose that $\alpha_1, \alpha_2, \ldots, \alpha_r$ are roots of $f(x)$. By the inductive hypothesis, $(x-\alpha_1)(x-\alpha_2)\ldots(x-\alpha_{r-1})$ is a factor of $f(x)$, so that we can write

$$f(x) = (x-\alpha_1)(x-\alpha_2)\ldots(x-\alpha_{r-1})g(x)$$

for some $g(x) \in R[x]$. We now evaluate at α_r to give

$$0 = (\alpha_r - \alpha_1)(\alpha_r - \alpha_2)\ldots(\alpha_r - \alpha_{r-1})g(\alpha_r).$$

Now $\alpha_r - \alpha_1, \ldots, \alpha_r - \alpha_{r-1}$ are all non-zero, and there are no proper divisors of zero in R, so that $g(\alpha_r) = 0$. It follows from Theorem 2.8.2 that $x - \alpha_r$ is a factor of $g(x)$. Hence, $(x-\alpha_1)\ldots(x-\alpha_{r-1})(x-\alpha_r)$ is a factor of $f(x)$. This completes the inductive step. □

The next result follows immediately from Corollary 2.8.3.

Corollary 2.8.4 Let R be an integral domain. Then a polynomial in $R[x]$ of degree $n \geqslant 0$ has at most n distinct roots. □

This result is not true if R is an arbitrary commutative ring. For example, in $\mathbb{Z}_8[x]$, the polynomial $x^2 - \bar{1}$ is of degree 2 and yet has four roots $\bar{1}, \bar{3}, \bar{5}, \bar{7}$.

We can improve on Corollary 2.8.4 when $R = \mathbb{C}$ in view of the next result.

Theorem 2.8.5 (The Fundamental Theorem of Algebra) Every polynomial with complex coefficients of positive degree has a root.

As we remarked in the introduction, the Fundamental Theorem of

2.8 Factorization of polynomials

Algebra is really a theorem of complex analysis. We shall take it as read here, and refer interested readers to any book on that subject for a proof–see, for example, *Complex Analysis*, subtitled *The Hitch Hiker's Guide to the Complex Plane*, pp. 185 and 233, by Ian Stewart and David Tall.

If we start with a polynomial in $\mathbb{C}[x]$ of positive degree, this will thus have a root and hence, by the Factor Theorem, it will have a factor $x - \alpha_1$, where $\alpha_1 \in \mathbb{C}$. We can do the same with what is left, assuming that it is of positive degree, and obtain a second factor $x - \alpha_2$, where $\alpha_2 \in \mathbb{C}$. Eventually we shall have expressed our polynomial in the form

$$\alpha(x-\alpha_1)(x-\alpha_2)\ldots(x-\alpha_n),$$

where $\alpha, \alpha_1, \alpha_2, \ldots, \alpha_n \in \mathbb{C}$. This describes the factorization of a polynomial in $\mathbb{C}[x]$ into irreducibles. In fact, we can say the following:

Corollary 2.8.6 Every polynomial in $\mathbb{C}[x]$ of positive degree is a product of linear factors (i.e. factors of degree 1), and the irreducible polynomials in $\mathbb{C}[x]$ are precisely the linear polynomials. □

We know that $\mathbb{C}[x]$ is a UFD (because it is a Euclidean domain), so that the factorization of a polynomial of positive degree into linear factors is essentially unique. In fact, if we express it in the form

$$\alpha(x-\alpha_1)(x-\alpha_2)\ldots(x-\alpha_n),$$

then it is unique apart from the order of the factors; α is the leading coefficient of the polynomial and $\alpha_1, \alpha_2, \ldots, \alpha_n$ are the roots. Thus a polynomial in $\mathbb{C}[x]$ of degree $n \geqslant 0$ has precisely n roots, provided that each root β is counted as many times as the factor $x - \beta$ occurs in the above factorization of the polynomial.

The Fundamental Theorem of Algebra will also deal with polynomials with real coefficients, which after all also belong to $\mathbb{C}[x]$. For consider the non-zero polynomial

$$f(x) = a_n x^n + a_{n-1} x^{n-1} + \ldots + a_1 x + a_0$$

in $\mathbb{R}[x]$. Then $f(x)$ can be regarded as belonging to $\mathbb{C}[x]$, and so will have precisely n complex roots. The real roots among these will give rise to real linear factors. Suppose that α is a non-real root. Then

$$a_n\alpha^n + a_{n-1}\alpha^{n-1} + \ldots + a_1\alpha + a_0 = 0.$$

If we now conjugate all terms then, because $a_0, a_1, \ldots, a_n$ are real, we have

$$a_n\bar{\alpha}^n + a_{n-1}\bar{\alpha}^{n-1} + \ldots + a_1\bar{\alpha} + a_0 = 0$$

and $\bar{\alpha}$, which is different from α, is also a root. Hence, $f(x)$ has a factor

$$(x-\alpha)(x-\bar{\alpha})$$
$$= x^2 - (\alpha + \bar{\alpha})x + \alpha\bar{\alpha}$$
$$= x^2 - 2R(\alpha)x + |\alpha|^2,$$

where $R(\alpha)$ denotes the real part of α and $|\alpha|$ denotes the modulus of α. This quadratic factor has real coefficients so, when $f(x)$ is divided by this factor, say

$$f(x) = q(x)(x^2 - 2R(\alpha)x + |\alpha|^2),$$

then $q(x)$ also belongs to $\mathbb{R}[x]$, and we can now consider the factorization of $q(x)$ in a similar way. This gives the following result:

Corollary 2.8.7 Every polynomial in $\mathbb{R}[x]$ of positive degree factorizes into irreducible real linear and quadratic factors, and every irreducible polynomial in $\mathbb{R}[x]$ is either linear (degree 1) or quadratic (degree 2). □

In fact, as in the case of polynomials in $\mathbb{C}[x]$, it is best to write a polynomial of positive degree in $\mathbb{R}[x]$ as a product of monic (i.e. leading coefficient 1) irreducible polynomials times a real number (its leading coefficient). Then this factorization is unique (apart from the order of the factors).

Readers are probably familiar with the test as to when a quadratic polynomial

$$ax^2 + bx + c$$

is irreducible in $\mathbb{R}[x]$. In $\mathbb{C}[x]$, we can factorize it to give

$$a\left(x - \frac{-b + \sqrt{(b^2 - 4ac)}}{2a}\right)\left(x - \frac{-b - \sqrt{(b^2 - 4ac)}}{2a}\right),$$

2.8 Factorization of polynomials

and these factors belong to $\mathbb{R}[x]$ if and only if $b^2 - 4ac \geq 0$. Since this factorization is unique in $\mathbb{C}[x]$, it follows that the quadratic polynomial $ax^2 + bx + c$ is irreducible in $\mathbb{R}[x]$ if and only if $b^2 - 4ac < 0$.

The situation is more complicated when we consider polynomials with coefficients in $\mathbb{Q}$. As an illustration, we consider a polynomial which arises from a famous problem of ancient Greek mathematics. Readers may recall how to bisect an angle using a straight edge and compass (and writing implement!) alone, but is there a similar construction for the trisection of an angle? Consider, for example, an angle of 60°. Put $\theta = 20°$. Then

$$\begin{aligned}
\tfrac{1}{2} = \cos 60° &= \cos 3\theta = \cos(2\theta + \theta) \\
&= \cos 2\theta \cos \theta - \sin 2\theta \sin \theta \\
&= (2\cos^2\theta - 1)\cos\theta - 2\sin^2\theta \cos\theta \\
&= 2\cos^3\theta - \cos\theta - 2\cos\theta(1 - \cos^2\theta) \\
&= 4\cos^3\theta - 3\cos\theta.
\end{aligned}$$

Thus $\cos 20°$ is a root of the polynomial $4x^3 - 3x - \tfrac{1}{2}$ in $\mathbb{R}$. One proof that it is impossible to trisect an angle of 60° using a straight edge and compass alone involves the assertion that this polynomial is irreducible in $\mathbb{Q}[x]$. We may as well multiply by 2, a unit of $\mathbb{Q}[x]$, to give the polynomial $8x^3 - 6x - 1$, with integer coefficients. It is easy to see that this polynomial is irreducible in $\mathbb{Z}[x]$. Since there is no question of taking out a constant factor (other than ± 1), a proper factorization must involve a linear factor, say

$$8x^3 - 6x - 1 = (ax + b)(cx^2 + dx + e),$$

where $a, b, c, d, e \in \mathbb{Z}$. Then $8 = ac$, $-1 = be$, so $a|8$, $b|(-1)$. Thus the only possible linear factors are

$$\pm x \pm 1, \quad \pm 2x \pm 1, \quad \pm 4x \pm 1, \quad \pm 8x \pm 1.$$

But then one of ± 1, $\pm \tfrac{1}{2}$, $\pm \tfrac{1}{4}$, $\pm \tfrac{1}{8}$ is a root of the polynomial, and it is easy to check that none of these is a root. Thus the polynomial in question is irreducible in $\mathbb{Z}[x]$. We shall see in Theorem 2.8.11 that we can thereby deduce that the polynomial is also irreducible in $\mathbb{Q}[x]$.

Clearly, when considering the factorization of polynomials in $\mathbb{Q}[x]$, we can do as we did in this example and multiply through the polynomial by an integer (a unit in $\mathbb{Q}[x]$) to clear the denominators among the coefficients and so start with a polynomial in $\mathbb{Z}[x]$.

We shall delay until the end of the section the more practical questions of how to factorize a polynomial in $\mathbb{Z}[x]$ into irreducible polynomials, and of how to tell when a polynomial in $\mathbb{Z}[x]$ is irreducible. We first deal with more general considerations, culminating in an important 'new lamps for old' theorem about UFDs. Thus, for the moment, R is an arbitrary UFD with field of fractions k (for example, $R = \mathbb{Z}$ and $k = \mathbb{Q}$). As we remarked for polynomials with rational coefficients, when considering the factorization of polynomials in $k[x]$, we can always multiply through by a non-zero element of R (a unit of $k[x]$) to give a polynomial in $R[x]$. Recall from Theorem 2.4.7 that, in a UFD, a finite non-empty set of elements possesses a GCD.

Definition 2.8.8 Let $f(x) \in R[x]$, $f(x) \neq 0$. The 'content' of $f(x)$ is the GCD of its non-zero coefficients. (Thus the content of $f(x)$ is uniquely determined up to units.) Further, $f(x)$ is said to be 'primitive' if its content is 1.

Thus, in $\mathbb{Z}[x]$, $8x^3 - 6x - 1$ is primitive whereas $8x^3 - 6x - 2$ has content 2. In view of Remark 2 following Definition 2.4.1, we note that, if $f(x) \in R[x]$, $f(x) \neq 0$, then we can take out its content d and write

$$f(x) = df^*(x),$$

where $f^*(x)$ is a primitive polynomial in $R[x]$. Thus, for example,

$$8x^3 - 6x - 2 = 2(4x^3 - 3x - 1).$$

Theorem 2.8.9 (Gauss's Lemma) In $R[x]$, the product of two primitive polynomials is primitive.

Proof Let

$$f(x) = a_0 + a_1 x + a_2 x^2 + \ldots,$$
$$g(x) = b_0 + b_1 x + b_2 x^2 + \ldots$$

be primitive polynomials in $R[x]$. We note that $f(x) \neq 0$, $g(x) \neq 0$, so

2.8 Factorization of polynomials

that $f(x)g(x) \neq 0$. Consider an irreducible element p of R. Since the GCD of the non-zero coefficients of $f(x)$ is 1, p cannot divide all the a_i's, so there exists r such that

$$p \mid a_0, a_1, \ldots, a_{r-1} \text{ but } p \nmid a_r.$$

Similarly, there exists s such that

$$p \mid b_0, b_1, \ldots, b_{s-1} \text{ but } p \nmid b_s.$$

The coefficient of x^{r+s} in $f(x)g(x)$ is

$$a_0 b_{r+s} + a_1 b_{r+s-1} + \ldots + a_{r-1} b_{s+1} + a_r b_s$$
$$+ a_{r+1} b_{s-1} + \ldots + a_{r+s} b_0.$$

Now

$$p \mid a_0 b_{r+s}, a_1 b_{r+s-1}, \ldots, a_{r-1} b_{s+1}, a_{r+1} b_{s-1}, \ldots, a_{r+s} b_0.$$

But R is a UFD, so, by Theorem 2.5.2(3), p is prime. Since $p \nmid a_r$, $p \nmid b_s$, it follows that $p \nmid a_r b_s$. It follows that p does not divide the coefficient of x^{r+s} in $f(x)g(x)$. Since p was an arbitrary irreducible element of R, it follows that the GCD of the non-zero coefficients of $f(x)g(x)$ must be 1, and $f(x)g(x)$ is thereby primitive. □

Theorem 2.8.10 Let $f(x) \in k[x]$, $f(x) \neq 0$. Then there exists $\alpha \in k$ and a primitive polynomial $f^*(x) \in R[x]$ such that $f(x) = \alpha f^*(x)$. Moreover, if $\alpha f^*(x) = \beta g^*(x)$, where α, β are non-zero elements of k and $f^*(x), g^*(x)$ are primitive polynomials in $R[x]$, then $\alpha = u\beta$ and $g^*(x) = u f^*(x)$ for some unit u of R.

For example, in $\mathbb{Q}[x]$,

$$\tfrac{3}{7}x^3 + \tfrac{9}{5}x = \tfrac{3}{35}(5x^3 + 21x),$$

where $\tfrac{3}{35} \in \mathbb{Q}$ and $5x^3 + 21x$ is a primitive polynomial in $\mathbb{Z}[x]$. Further, the only other way in which the polynomial can be so expressed is to write

$$\tfrac{3}{7}x^3 + \tfrac{9}{5}x = -\tfrac{3}{35}(-5x^3 - 21x),$$

because the only units of $\mathbb{Z}$ are ± 1.

70   *Factorization*

Proof Write

$$f(x) = \frac{a_0}{b_0} + \frac{a_1}{b_1}x + \frac{a_2}{b_2}x^2 + \ldots + \frac{a_n}{b_n}x^n,$$

where the a_i, b_i belong to R with the b_i non-zero. Put $b = b_0 b_1 b_2 \ldots b_n$. Then we can write

$$f(x) = \frac{1}{b}\bar{f}(x),$$

where $\bar{f}(x) \in R[x]$. Let the GCD of the non-zero coefficients of $\bar{f}(x)$ be d. Then we can write $\bar{f}(x) = df^*(x)$, where $f^*(x)$ is a primitive polynomial in $R[x]$. Thus

$$f(x) = \frac{d}{b}f^*(x),$$

which proves the 'existence' part.

Now suppose that

$$\alpha f^*(x) = \beta g^*(x),$$

where α, β are non-zero elements of k and $f^*(x), g^*(x)$ are primitive polynomials in $R[x]$. Put $\alpha = a_1/b_1$, $\beta = a_2/b_2$, where a_1, a_2, b_1, b_2 are non-zero elements of R. Then

$$a_1 b_2 f^*(x) = a_2 b_1 g^*(x).$$

By Remark 3 following Definition 2.4.1, the GCD of the non-zero coefficients of the left-hand side is $a_1 b_2$, and that of the non-zero coefficients of the right-hand side is $a_2 b_1$. Hence, $a_1 b_2 = u a_2 b_1$ for some unit u of R, whence $\alpha = u\beta$. It follows also that $g^*(x) = u f^*(x)$. □

Theorem 2.8.11 Let $f(x)$ be a polynomial in $R[x]$ of positive degree, and suppose that

$$f(x) = g(x)h(x),$$

where $g(x), h(x) \in k[x]$. Then there exist polynomials $g_1(x), h_1(x) \in R[x]$ such that $\deg g_1(x) = \deg g(x)$, $\deg h_1(x) = \deg h(x)$ and such that

$$f(x) = g_1(x)h_1(x).$$

2.8 Factorization of polynomials

(Thus, if $f(x)$ has been factorized in $k[x]$, it can be refactorized in $R[x]$ with factors of the same degree as the original factors.) Hence, if $f(x)$ is irreducible in $R[x]$, it is also irreducible in $k[x]$.

Remark It is not true that, if a polynomial $f(x)$ of $R[x]$ is irreducible in $k[x]$, then it is irreducible in $R[x]$. For example, $2x+4$ is irreducible in $\mathbb{Q}[x]$, but it can be factorized as $2(x+2)$ in $\mathbb{Z}[x]$, and neither factor is a unit of $\mathbb{Z}[x]$. What is true is that, if $f(x) \in R[x]$ is primitive and irreducible in $k[x]$, then it is irreducible in $R[x]$. For suppose that $f(x)$ is primitive and irreducible in $k[x]$, and write $f(x) = g(x)h(x)$, where $g(x), h(x) \in R[x]$. Then $g(x), h(x) \in k[x]$, so one of them is a unit of $k[x]$, say $g(x)$, i.e $g(x)$ is of degree zero. Thus $g(x) \in R$ and divides the non-zero coefficients of $f(x)$. Since $f(x)$ is primitive, $g(x)$ must thus be a unit of R and so of $R[x]$.

Proof of Theorem 2.8.11 We can write $f(x) = df^*(x)$, where $d \in R$ is the content of $f(x)$ and $f^*(x)$ is a primitive polynomial in $R[x]$. By Theorem 2.8.10, we can also write

$$g(x) = \alpha g^*(x), \ h(x) = \beta h^*(x),$$

where $\alpha, \beta \in k$ and $g^*(x), h^*(x)$ are primitive polynomials in $R[x]$. Then

$$df^*(x) = \alpha \beta g^*(x) h^*(x).$$

By Gauss's Lemma (Theorem 2.8.9), $g^*(x)h^*(x)$ is primitive so, by the uniqueness part of Theorem 2.8.10, $\alpha\beta = ud$ for some unit u of R. Thus $\alpha\beta \in R$. Also,

$$f(x) = (\alpha\beta g^*(x)) h^*(x).$$

Put $g_1(x) = \alpha\beta g^*(x)$, $h_1(x) = h^*(x)$. Then $g_1(x), h_1(x) \in R[x]$, $\deg g_1(x) = \deg g(x)$, $\deg h_1(x) = \deg h(x)$ and $f(x) = g_1(x)h_1(x)$.

Finally, suppose that $f(x)$ is irreducible in $R[x]$ and that $f(x) = g(x)h(x)$, where $g(x), h(x) \in k[x]$. We can find $g_1(x), h_1(x) \in R[x]$ as described in the previous part. Since $f(x)$ is irreducible in $R[x]$, this means that one of $g_1(x), h_1(x)$ is a unit of $R[x]$, and so is of degree zero. Hence, one of $g(x), h(x)$ is of degree zero, and so is a unit of $k[x]$ because k is a field. Hence, $f(x)$ is irreducible in $k[x]$. □

72   Factorization

We can now prove an important 'new lamps for old' or, rather, 'new UFDs for old' theorem.

Theorem 2.8.12   If R is a UFD, then so is the polynomial ring $R[x]$.

It will follow by repeated application of this result that, if R is a UFD, so is the polynomial ring $R[x_1,\ldots,x_n]$ in n independent variables over R. Thus, for example, $\mathbb{Z}[x_1,\ldots,x_n]$ is a UFD. Similarly, every field k is a UFD, so the polynomial ring $k[x_1,\ldots,x_n]$ is a UFD.

Proof of Theorem 2.8.12   Let R be a UFD, and consider $f(x) \in R[x]$, not zero and not a unit. If we denote the content of $f(x)$ by d, then $f(x) = df^*(x)$, where $d \in R$ and $f^*(x)$ is primitive in $R[x]$. We now factorize d, $f^*(x)$ separately into irreducibles in $R[x]$. If d is a unit of R (and so of $R[x]$), then we need only consider $f^*(x)$. If $f^*(x)$ is a unit of $R[x]$ (when it will also be a unit of R), we need only consider the factorization of d. Of course, they cannot both be units because $f(x)$ is not a unit. Since R is a UFD, we can factorize d (if a non-unit) into irreducibles in R, to give $d = p_1 p_2 \ldots p_r$. Further, the p_i are also irreducible in $R[x]$ since the only way of factoring p_i is to take factors in R. Now consider $f^*(x)$ (if a non-unit). If $f^*(x)$ is irreducible in $R[x]$, fine! Otherwise, since it is primitive, we can write $f^*(x) = g(x)h(x)$, where $g(x), h(x) \in R[x]$, $\deg g(x) \geq 1$, $\deg h(x) \geq 1$. Also, $g(x), h(x)$ are primitive because $f^*(x)$ is. If $g(x)$ is not irreducible, then it can be further factorized, as can $h(x)$. We can continue in this way. Because the degrees of the factors are going down with each factorization, and $f^*(x)$ has a definite degree to begin with, this process must stop. When this happens, $f^*(x)$ will have been factorized into irreducibles.

To prove uniqueness, suppose that

$$p_1 \ldots p_r f_1(x) \ldots f_s(x) = q_1 \ldots q_t g_1(x) \ldots g_w(x),$$

where the p_i, q_j are in R and the $f_i(x)$, $g_j(x)$ are in $R[x]$ and are of positive degree, and all are irreducible elements of $R[x]$. Then the p_i, q_j will be irreducible elements of R. The $f_i(x), g_j(x)$ are primitive because they are irreducible, so $f_1(x)\ldots f_s(x)$ and $g_1(x)\ldots g_w(x)$ are primitive by Gauss's Lemma (Theorem 2.8.9). (Note that, from

2.8 Factorization of polynomials

consideration of degrees, $s=0$ if and only if $w=0$.) Hence, by the uniqueness part of Theorem 2.8.10,

$$f_1(x)\ldots f_s(x) = ug_1(x)\ldots g_w(x)$$

for some unit u of R. We are trying to pair off the p_i's and the q_j's and the $f_i(x)$'s and the $g_j(x)$'s to give associate pairs, so there is no harm in replacing $g_1(x)$ by $ug_1(x)$ and q_1 by $u^{-1}q_1$. When this is done, we can say that

$$f_1(x)\ldots f_s(x) = g_1(x)\ldots g_w(x),$$

whence also

$$p_1 \ldots p_r = q_1 \ldots q_t.$$

From the uniqueness of factorization in R, $r=t$ and the p_i, q_j can be paired off into associate pairs in R and so in $R[x]$. (Note that $r=0$ if and only if $t=0$.)

It remains to consider the equation

$$f_1(x)\ldots f_s(x) = g_1(x)\ldots g_w(x).$$

This next step provides the only redeeming feature in an otherwise tedious argument. Denote by k the field of fractions of R. By Theorem 2.8.11, the $f_i(x), g_j(x)$ are also irreducible in $k[x]$, and $k[x]$ is a UFD by Theorems 2.3.2 and 2.6.1. Hence, $s=w$ and the $f_i(x), g_j(x)$ can be paired off to give associate pairs in $k[x]$. Suppose that $f_1(x), g_1(x)$ are associates in $k[x]$. Then

$$f_1(x) = \beta g_1(x),$$

where β is a non-zero element of k. It now follows from Theorem 2.8.10 (with $\alpha=1$) that $g_1(x) = uf_1(x)$ for some unit u of R, and $f_1(x)$, $g_1(x)$ are actually associates in $R[x]$. This completes the proof of the uniqueness. □

We now return to the questions of how to factorize a polynomial in $\mathbb{Z}[x]$ into irreducible factors and of how to tell when a polynomial in $\mathbb{Z}[x]$ is irreducible. In view of Theorem 2.8.12, we now know that $\mathbb{Z}[x]$ is a UFD. If we return to the example of the polynomial $8x^3 - 6x - 1$ in $\mathbb{Z}[x]$, we showed that this is irreducible in $\mathbb{Z}[x]$, and therefore in $\mathbb{Q}[x]$ from Theorem 2.8.11, by showing that it has no linear factors, and this we did by eliminating possible rational roots.

The connection between linear factors and rational roots is a close one. Suppose that R is a general integral domain with field of fractions k. If the polynomial $f(x) \in R[x]$ has a linear factor $ax+b$ (with $a, b \in R$ and $a \neq 0$), then $-b/a$ is a root of $f(x)$ in k. Also, consideration of the leading coefficient and 'constant term' shows that a divides the leading coefficient of $f(x)$ in R and b divides its constant term. Now let R be a UFD, and suppose conversely that $f(x)$ has a root c/d say in k (with $c, d \in R$, $d \neq 0$). Now c, d have a GCD which can be cancelled, so we can suppose that c, d are coprime. We say in this case that the fraction c/d is 'in lowest terms'. By Theorem 2.8.2, we can write

$$f(x) = \left(x - \frac{c}{d}\right) g(x)$$

for some $g(x) \in k[x]$. The proof of Theorem 2.8.11 now shows that we can transfer $1/d$ from the first factor to the second to give

$$f(x) = (dx - c)\left(\frac{1}{d} g(x)\right)$$

and that the two factors are now in $R[x]$. (Note that $dx - c$ is a primitive polynomial.) Thus $f(x)$ has a linear factor in $R[x]$ (note that $R[x]$ is a UFD), and d divides the leading coefficient of $f(x)$ and c divides its constant term.

It seems worthwhile to include a direct proof of this last result, without appeal to the proof of Theorem 2.8.11. This requires the stronger hypothesis that R is a Euclidean domain, but, since the usual application is when $R = \mathbb{Z}$, this presents no problem.

Theorem 2.8.13 Let R be a Euclidean domain with field of fractions k, let

$$f(x) = a_n x^n + a_{n-1} x^{n-1} + \ldots + a_1 x + a_0 \in R[x]$$

(with $a_0, a_1, \ldots, a_n \in R$) and let c/d be a root of $f(x)$ in k (with $c, d \in R$, $d \neq 0$ and c, d coprime). Then $c | a_0$ and $d | a_n$ in R.

2.8 Factorization of polynomials

Proof We have

$$a_n\left(\frac{c}{d}\right)^n + a_{n-1}\left(\frac{c}{d}\right)^{n-1} + \ldots + a_0 = 0,$$

so that, if we multiply by d^n, we obtain

$$a_n c^n + a_{n-1} c^{n-1} d + \ldots + a_0 d^n = 0.$$

Thus $c(a_n c^{n-1} + \ldots + a_1 d^{n-1}) = -a_0 d^n$, so that $c | a_0 d^n$, and similarly $d | a_n c^n$. But c, d are coprime and R is a Euclidean domain. Successive application of Theorem 2.4.6 shows that $c | a_0 d^{n-1}, a_0 d^{n-2}, \ldots,$ and eventually that $c | a_0$. Similarly, $d | a_n$. □

The above helps to find linear factors of polynomials in $\mathbb{Z}[x]$, or to show that our polynomial has none. This will be sufficient to show that a polynomial of degree 3 or less is irreducible in $\mathbb{Z}[x]$ (if indeed it is), but it will not cope with polynomials of degree greater than 3, which may have non-linear factors. It is helpful to have criteria which tell us that certain polynomials are irreducible in $\mathbb{Z}[x]$ (or in $\mathbb{Q}[x]$). One such was provided by Eisenstein, who was a pupil of Gauss. Gauss placed Eisenstein in the top three mathematicians of all time. Even Gauss could be wrong! We may as well continue with a general UFD R rather than just $\mathbb{Z}$, although our applications will invariably be when $R = \mathbb{Z}$.

Theorem 2.8.14 (Eisenstein's Irreducibility Criterion) Let R be a UFD, let

$$f(x) = a_n x^n + a_{n-1} x^{n-1} + \ldots + a_1 x + a_0$$

be a polynomial of positive degree in $R[x]$, and suppose that there is an irreducible element p of R such that

$$p \nmid a_n, p | a_{n-1}, a_{n-2}, \ldots, a_0, p^2 \nmid a_0.$$

Then $f(x)$ is irreducible in $k[x]$. Similarly, if there is an irreducible element q of R such that

$$q \nmid a_0, q | a_1, a_2, \ldots, a_n, q^2 \nmid a_n,$$

then again $f(x)$ is irreducible in $k[x]$.

Before we prove the criterion, we remark that the conditions given do not allow us to conclude that $f(x)$ is irreducible in $R[x]$; for that the polynomial needs to be primitive as well. For example, the polynomial $2x+6$ is not irreducible in $\mathbb{Z}[x]$ even though $3 \nmid 2$, $3|6$ and $3^2 \nmid 6$. Note that 3 is an irreducible element of $\mathbb{Z}$ because it is a prime number.

As a result of this criterion, we can immediately provide irreducible polynomials in $\mathbb{Q}[x]$ (and in $\mathbb{Z}[x]$) of every given positive degree, contrasting with $\mathbb{C}[x]$, where they are all linear, and $\mathbb{R}[x]$, where they are linear or quadratic. Thus, for an irreducible polynomial in $\mathbb{Q}[x]$ (and in $\mathbb{Z}[x]$ because it is primitive) of degree n, where n is any positive integer, we can take $x^n + p$, where p is any prime number, or $x^n - p$, or $x^n + px^{n-1} + \ldots + px + p$, or any number of others. In the case of $x^n + p$, we have $p \nmid 1, p|0, \ldots, 0, p$ and $p^2 \nmid p$.

Proof of Eisenstein's Irreducibility Criterion We shall just prove the result involving the irreducible element p; the proof in the case of q simply means working from the other end of the polynomial. We start with a factorization of $f(x)$ in $k[x]$ into two factors. By Theorem 2.8.11, there will then be a factorization

$$f(x) = (b_m x^m + b_{m-1} x^{m-1} + \ldots + b_0)$$
$$\times (c_{n-m} x^{n-m} + c_{n-m-1} x^{n-m-1} + \ldots + c_0),$$

where the b_i, c_j belong to R and the two factors have the same degrees as the original factors in $k[x]$. Now $a_0 = b_0 c_0$ and $p|a_0$, $p^2 \nmid a_0$. Also, by Theorem 2.5.2(3), p is prime. Hence, p divides one of b_0, c_0 but not both, say $p \nmid b_0$, $p|c_0$. Also, $a_n = b_m c_{n-m}$ and $p \nmid a_n$, so that $p \nmid c_{n-m}$. Thus there exists r with $1 \leqslant r \leqslant n-m$ such that $p|c_0, c_1, \ldots, c_{r-1}$ but $p \nmid c_r$. Now

$$a_r = b_0 c_r + b_1 c_{r-1} + \ldots + b_r c_0$$

and $p|b_r c_0, b_{r-1} c_1, \ldots, b_1 c_{r-1}$ yet $p \nmid b_0 c_r$ (by Theorem 2.5.2(3) again). Thus $p \nmid a_r$. Since the only a_i not divisible by p is a_n, this means that $r = n$, so that $m = 0$ and the first factor in the original factorization of $f(x)$ in $k[x]$ has degree zero, and so is a unit of $k[x]$. Hence, $f(x)$ is irreducible in $k[x]$. □

2.8 Factorization of polynomials

Example Eisenstein's Irreducibility Criterion is not always applied in a straightforward way, as we shall see. Let p be a prime number in $\mathbb{Z}$, and consider the polynomial $x^p - 1$ in $\mathbb{Q}[x]$. Its complex roots are the pth roots of unity. It can be factorized in $\mathbb{Q}[x]$ as

$$x^p - 1 = (x-1)(x^{p-1} + x^{p-2} + \ldots + x + 1).$$

We assert that this is the factorization of $x^p - 1$ into monic irreducible factors in $\mathbb{Q}[x]$ (and also in $\mathbb{Z}[x]$, since the factors are primitive). Certainly $x - 1$ is irreducible. We look at the so-called *cyclotomic* ('*circle-dividing*') *polynomial*

$$x^{p-1} + x^{p-2} + \ldots + x + 1 = \phi_p(x)$$

(say). (This plays an important rôle in Gauss's theory of constructible regular polygons.) Eisenstein's Irreducibility Criterion does not apply to the polynomial as it stands. But suppose that we change the variable by putting $x = y + 1$. To be precise, what we are doing is mapping from the polynomial ring $\mathbb{Q}[x]$ to the polynomial ring $\mathbb{Q}[y]$ by mapping an arbitrary polynomial $f(x)$ to $f(y+1)$. It is easy to see that this mapping is a homomorphism. If we apply this homomorphism to the factorization

$$x^p - 1 = (x-1)\phi_p(x)$$

already given, we obtain

$$(y+1)^p - 1 = y\phi_p(y+1),$$

so that, if we expand the left-hand side using the binomial theorem and cancel y, we obtain

$$\phi_p(y+1) = y^{p-1} + \binom{p}{1}y^{p-2} + \binom{p}{2}y^{p-3} + \ldots + \binom{p}{p-1},$$

where, for $1 \leq k \leq p-1$,

$$\binom{p}{k} = \frac{p!}{k!(p-k)!}$$

denotes the familiar binomial coefficient. Now these coefficients are all integers divisible by p (see Exercise 2.5.2), and

$$\binom{p}{p-1} = p$$

and so is not divisible by p^2. It follows from Eisenstein's Irreducibility Criterion that $\phi_p(y+1)$ is irreducible in $\mathbb{Q}[y]$. Any factorization $\phi_p(x) = f(x)g(x)$ in $\mathbb{Q}[x]$ will give a factorization $\phi_p(y+1) = f(y+1)g(y+1)$ in $\mathbb{Q}[y]$, using the homomorphism $\mathbb{Q}[x] \to \mathbb{Q}[y]$ described above, and $f(y+1), g(y+1)$ are polynomials in $\mathbb{Q}[y]$ of the same degrees as $f(x), g(x)$ in $\mathbb{Q}[x]$, respectively. Since $\phi_p(y+1)$ is irreducible in $\mathbb{Q}[y]$, one of $f(y+1), g(y+1)$ must be a unit, and so is of degree zero. Hence, one of $f(x), g(x)$ must be of degree zero in $\mathbb{Q}[x]$, and so is a unit. Hence, $\phi_p(x)$ is irreducible in $\mathbb{Q}[x]$.

In the above example, we have used the technique of substituting $x = y+1$ to transform the cyclotomic polynomial into a form to which Eisenstein's Irreducibility Criterion can be applied. This trick obviously can be applied more generally. Thus we may wish to show that a given polynomial $f(x)$ in $R[x]$ (R a UFD) is irreducible in $k[x]$, k being the field of fractions of R. If we can find $a \in R$ such that, when we put $x = y+a$, we obtain a polynomial $f(y+a)$ which can be shown to be irreducible in $k[y]$, say by Eisenstein's Irreducibility Criterion, it follows as above (with a in place of 1) that the original polynomial $f(x)$ is irreducible in $k[x]$. The art will be to choose a suitable element a. Further examples where this technique is useful may be found in the exercises at the end of this section.

We are still left with the question of how to factorize a polynomial in $\mathbb{Z}[x]$ into its irreducible factors. This is often a laborious task, although algorithms for this do exist. We describe one such, due to Kronecker. We shall need one tool, which we shall prove first. Its use will be illustrated shortly.

Theorem 2.8.15 (Lagrange's Interpolation Formula) Let k be a field, let $a_0, a_1, \ldots, a_m$ be distinct elements of k and let $b_0, b_1, \ldots, b_m \in k$. Then there is a unique polynomial $g(x)$ in $k[x]$ of degree $\leq m$ such that $g(a_i) = b_i$ for $0 \leq i \leq m$, given by

$$g(x) = \sum_{r=0}^{m} b_r \frac{(x-a_0)\ldots(x-a_{r-1})(x-a_{r+1})\ldots(x-a_m)}{(a_r-a_0)\ldots(a_r-a_{r-1})(a_r-a_{r+1})\ldots(a_r-a_m)}.$$

Proof It is a matter of simple verification that the polynomial given by the formula has the properties stated. To see that this is the only

2.8 Factorization of polynomials

such polynomial, consider two such, $g_1(x)$ and $g_2(x)$. Then $a_0, a_1, \ldots, a_m$ are $m+1$ distinct roots of the polynomial $g_1(x) - g_2(x)$, yet $\deg(g_1(x) - g_2(x)) \leq m$, so, by Corollary 2.8.4, $g_1(x) - g_2(x) = 0$, i.e. $g_1(x) = g_2(x)$. Thus the polynomial is unique. □

We now describe Kronecker's method of factorizing a polynomial in $\mathbb{Z}[x]$. In fact we can describe it quite generally, so we start with a non-zero polynomial $f(x) \in R[x]$, where R is an arbitrary UFD. Then $R[x]$ is also a UFD, and we look for a factorization $f(x) = g(x)h(x)$ with $g(x), h(x) \in R[x]$. Put $n = \deg f(x)$, and suppose that $\deg g(x) \leq \deg h(x)$. Then $\deg g(x) \leq n/2$; in fact $\deg g(x) \leq m$, where $m = n/2$ if n is even and $(n-1)/2$ if n is odd. By Corollary 2.8.4, $f(x)$ has at most n distinct roots, so we can find $m+1$ distinct elements $a_0, a_1, \ldots, a_m$ of R such that $f(a_0), f(a_1), \ldots, f(a_m)$ are not zero. Now $f(a_i) = g(a_i)h(a_i)$ for $0 \leq i \leq m$, so that $g(a_i) | f(a_i)$. Now R is a UFD, so we can factorize $f(a_i)$ into irreducibles in R, and thereby find its possible factors. If b_i is one such, then a possible value of $g(a_i)$ is b_i. Thus we obtain possible values $g(a_i) = b_i$ for $0 \leq i \leq m$. By Theorem 2.8.15, there is only one polynomial with these properties of degree $\leq m$ in $k[x]$, where k is the field of fractions of R, and Lagrange's Interpolation Formula will give this polynomial. If it turns out that this polynomial belongs to $R[x]$, then it is a candidate to be a factor of $f(x)$ in $R[x]$, and we can use the division algorithm to divide $f(x)$ by it in $k[x]$ to see if the quotient is in $R[x]$ and the remainder is zero. If so, then we have found a proper factorization of $f(x)$ in $R[x]$.

The method is best illustrated by an example. If nothing else, the example may illustrate the difficulty encountered in attempting to factorize even quite a simple polynomial in $\mathbb{Z}[x]$ into its irreducible factors.

Example 2.8.16 Factorize

$$2x^5 + 3x^4 + 3x^3 + 3x^2 + 8x + 6$$

into irreducible factors in $\mathbb{Z}[x]$.

Solution 1 (*not using Kronecker's method*) We first check for a possible linear factor $ax + b$ with $a, b \in \mathbb{Z}$. Then $a | 2, b | 6$, so $a = \pm 1$ or

± 2 and $b = \pm 1, \pm 2, \pm 3, \pm 6$, which would mean one of $\pm 1, \pm 2, \pm 3, \pm 6, \pm \frac{1}{2}, \pm \frac{3}{2}$ being a rational root. We leave readers to verify that none of these is a root, so that there is no linear factor. A quadratic factor (taking a positive leading coefficient) will have to be of one of the forms

$$x^2 + cx + d \quad \text{or} \quad 2x^2 + cx + d$$

with $c \in \mathbb{Z}$ and $d = \pm 1, \pm 2, \pm 3, \pm 6$. Try, for example, $x^2 + cx + 1$. Thus we have

$$2x^5 + 3x^4 + 3x^3 + 3x^2 + 8x + 6$$
$$= (x^2 + cx + 1)(2x^3 + rx^2 + sx + 6)$$

for some $r, s \in \mathbb{Z}$. Equating coefficients, we have the equations

$$r + 2c = 3$$
$$s + cr + 2 = 3$$
$$6 + cs + r = 3$$
$$6c + s = 8.$$

Thus $r = 3 - 2c$, $s = 8 - 6c$, so that

$$6 + c(8 - 6c) + (3 - 2c) = 3,$$
$$6c^2 - 6c - 6 = 0,$$

which has no solution in integers. It is now a case of trying all the possibilities, until the factorization

$$(x^2 + 2x + 2)(2x^3 - x^2 + x + 3)$$

is arrived at. Since the polynomial has no linear factor in $\mathbb{Z}[x]$, these factors are irreducible.

Solution 2 (using Kronecker's method) As in Solution 1, it is best first to eliminate the possibility of linear factors. Denote the given polynomial by $f(x)$, and write $f(x) = g(x)h(x)$ with $g(x), h(x) \in \mathbb{Z}[x]$ and $\deg g(x) = 2$. It is best to evaluate at values which give the smallest possible integers with the fewest factors, for obvious reasons, so we choose $0, -1, -2$. Now $f(0) = 6$, $f(-1) = -1$, $f(-2) = -38$. Thus $g(0)|6$, $g(-1)|(-1)$, $g(-2)|(-38)$, so that

2.8 Factorization of polynomials

$$g(0) = \pm 1, \pm 2, \pm 3 \text{ or } \pm 6,$$
$$g(-1) = \pm 1,$$
$$g(-2) = \pm 1, \pm 2, \pm 19 \text{ or } \pm 38.$$

It is now a case of trial and error. Lagrange's Interpolation Formula tells us that the polynomial $g(x)$ in $\mathbb{Q}[x]$ of degree ≤ 2 such that $g(0) = b_0$, $g(-1) = b_1$, $g(-2) = b_2$ is given by

$$g(x) = b_0 \frac{(x+1)(x+2)}{(0+1)(0+2)} + b_1 \frac{x(x+2)}{(-1-0)(-1+2)}$$
$$+ b_2 \frac{x(x+1)}{(-2-0)(-2+1)}$$
$$= \tfrac{1}{2} b_0 (x+1)(x+2) - b_1 x(x+2) + \tfrac{1}{2} b_2 x(x+1).$$

The values $b_0 = b_1 = b_2 = 1$ give the polynomial 1, which tells us nothing. The values $b_0 = b_1 = 1$, $b_2 = -1$ give the polynomial $-x^2 - x + 1$. It can be verified by long division that this does not divide the given polynomial. Similar things happen with other possibilities for b_0, b_1, b_2 until we stumble upon $b_0 = 2, b_1 = 1, b_2 = 2$, which gives the polynomial $x^2 + 2x + 2$, which divides into the given polynomial to give the quotient $2x^3 - x^2 + x + 3$ as in Solution 1.

Perhaps Kronecker's method is not very satisfactory as a practicable procedure for factorizing a polynomial in $\mathbb{Z}[x]$. But it does bring out how the problem of finding the factors of a polynomial in $\mathbb{Z}[x]$ reduces to that of finding factors of integers. In practice, of course, computers are used. A lot of work has been done in this area recently by A.K. Lenstra, H.W. Lenstra and L. Lovász to produce different algorithms and to estimate the time taken by a computer to do the factorization. (See their research paper entitled 'Factoring Polynomials with Rational Coefficients' in *Mathematische Annalen*.) The computational complexity of factorization in $\mathbb{Z}[x]$ is of the same order as that of factorization in $\mathbb{Z}$.

Exercises 2.8

1. Factorize $x^4 + 15$ into irreducible factors (a) in $\mathbb{C}[x]$, (b) in $\mathbb{R}[x]$, (c) in $\mathbb{Q}[x]$.

2. In the case of each of the following polynomials, determine whether it is irreducible (a) in $\mathbb{Q}[x]$, (b) in $\mathbb{Z}[x]$:

   (i) $x^3 + 2x^2 + 2x + 4$
   (ii) $x^3 + 6x^2 + 5x + 25$
   (iii) $3x^{10} + 6x^7 + 12x^3 + 6$
   (iv) $x^5 + x^4 + x^3 + x^2 + x + 1$
   (v) $3x^5 + 3x^4 + 9x^3 + 27x^2 + 18x + 1$
   (vi) $x^5 + 5x^2 + 4$
   (vii) $3x^7 + 7x^5 + 14x^3 + 7x + 16$
   (viii) $12x^{10} - 24x^7 + 18x^3 - 5$
   (ix) $2x^5 - 5x^4 - 5x + 36$.

3. Write down all the polynomials of degree 3 in $\mathbb{Z}_2[x]$ and factorize each one into its irreducible factors.

4. Let $p > 1$ be an integer. For

   $$f(x) = a_0 + a_1 x + a_2 x^2 + \ldots \in \mathbb{Z}[x],$$

   where $a_0, a_1, a_2, \ldots \in \mathbb{Z}$, put

   $$\bar{f}(x) = \bar{a}_0 + \bar{a}_1 x + \bar{a}_2 x^2 + \ldots \in \mathbb{Z}_p[x].$$

   Suppose that $f(x)$ is monic (i.e. it has leading coefficient 1). Show that, if $\bar{f}(x)$ is irreducible in $\mathbb{Z}_p[x]$, then $f(x)$ is irreducible in $\mathbb{Z}[x]$. Is the converse true?

   Show that $x^3 + ax^2 + bx + c$, with $a, b, c \in \mathbb{Z}$, is irreducible in $\mathbb{Z}[x]$ if $a + b$ and c are both odd.

5. Show that $\mathbb{Z}[x]$ and the ring $k[x, y]$ of polynomials in two independent variables over a field k are UFDs but not Euclidean domains.

6. Factorize $3x^5 + 4x^4 + 6x^3 - 3x - 2$ into irreducible factors in $\mathbb{Z}[x]$.

2.9

An application of UFDs to determinantal identities

We recall from Theorem 2.8.12 that, when R is a UFD, then so is the polynomial ring $R[x_1, \ldots, x_n]$ in n independent variables $x_1, \ldots, x_n$.

2.9 Application of UFDs to determinantal identities

We shall show in this final section how this fact may be exploited to prove various determinantal identities.

First a few general remarks about polynomials in many variables. Let R be an integral domain and let $x_1,\ldots,x_n$ be independent variables over R. Let r be a non-negative integer. A *non-zero* polynomial in $R[x_1,\ldots,x_n]$ of the form

$$f(x_1,\ldots,x_n) = \sum_{i_1+\ldots+i_n=r} a_{i_1\ldots i_n} x_1^{i_1} \ldots x_n^{i_n}$$

where the $a_{i_1\ldots i_n}$ belong to R, is said to be *homogeneous of degree* r. Thus, for example, in $\mathbb{Z}[x,y,z]$,

$$x^2y^3 + 2x^3yz + 5yz^4 + x^4z$$

is homogeneous of degree 5 because, in each summand of the polynomial, the sum of the powers of the variables is 5. However, $x^2y + xyz^2$ is not homogeneous. If $f(x_1,\ldots,x_n)$ is homogeneous of degree r and $g(x_1,\ldots,x_n)$ (assumed non-zero) is homogeneous of degree s, say

$$g(x_1,\ldots,x_n) = \sum_{j_1+\ldots+j_n=s} b_{j_1\ldots j_n} x_1^{j_1} \ldots x_n^{j_n},$$

then the product $f(x_1,\ldots,x_n)g(x_1,\ldots,x_n)$ is non-zero (because $R[x_1,\ldots,x_n]$ is an integral domain) and

$$f(x_1,\ldots,x_n)g(x_1,\ldots,x_n)$$
$$= \sum_{k_1+\ldots+k_n=r+s} \left(\sum_{i_\alpha+j_\alpha=k_\alpha} a_{i_1\ldots i_n} b_{j_1\ldots j_n} \right) x_1^{k_1} \ldots x_n^{k_n}.$$

In other words, the product of two homogeneous polynomials in $R[x_1,\ldots,x_n]$, one of degree r and the other of degree s, is a homogeneous polynomial of degree $r+s$. Note that the zero polynomial is excluded from this statement.

A general non-zero polynomial $f(x_1,\ldots,x_n)$ in $R[x_1,\ldots,x_n]$ is a sum of its homogeneous components, so that we can write

$$f(x_1,\ldots,x_n) = \sum_r f_r(x_1,\ldots,x_n),$$

where $f_r(x_1,\ldots,x_n)$ is the homogeneous component of $f(x_1,\ldots,x_n)$ of degree r. For example, $x^2y + xyz^2$ in $\mathbb{Z}[x,y,z]$ has homogeneous components x^2y, xyz^2 of degrees 3, 4, respectively. The largest non-negative integer r for which $f(x_1,\ldots,x_n)$ has a homogeneous component of degree r is called the *degree* of $f(x_1,\ldots,x_n)$. It is clear that this is the familiar degree in the case of a polynomial in one variable.

Let

$$g(x_1,\ldots,x_n) = \sum_s g_s(x_1,\ldots,x_n)$$

be another non-zero polynomial in $R[x_1,\ldots,x_n]$, expressed in terms of its homogeneous components. Then the homogeneous component of the product $f(x_1,\ldots,x_n)g(x_1,\ldots,x_n)$ of degree t is

$$\sum_{r+s=t} f_r(x_1,\ldots,x_n)g_s(x_1,\ldots,x_n).$$

Two things should be clear from this: (1) if either $f(x_1,\ldots,x_n)$ or $g(x_1,\ldots,x_n)$ is not homogeneous, then neither is their product, (2) the degree of the product is the sum of the degrees.

With these observations, we can now evaluate various determinantal identities. As an illustration, we consider *Vandermonde's determinant*

$$D = \begin{vmatrix} a_1^{n-1} & a_2^{n-1} & \cdots & a_n^{n-1} \\ a_1^{n-2} & a_2^{n-2} & \cdots & a_n^{n-2} \\ \vdots & \vdots & \cdots & \vdots \\ a_1 & a_2 & \cdots & a_n \\ 1 & 1 & \cdots & 1 \end{vmatrix}.$$

We shall take $a_1,\ldots,a_n$ as elements of an arbitrary commutative ring R, with 1 the identity element of R, although classically R would be the field $\mathbb{R}$ of real numbers. Note that we do not require R even to be an integral domain. We wish to evaluate this determinant. We consider instead the determinant

$$\Delta = \begin{vmatrix} x_1^{n-1} & x_2^{n-1} & \cdots & x_n^{n-1} \\ x_1^{n-2} & x_2^{n-2} & \cdots & x_n^{n-2} \\ \vdots & \vdots & \cdots & \vdots \\ x_1 & x_2 & \cdots & x_n \\ 1 & 1 & \cdots & 1 \end{vmatrix},$$

2.9 Application of UFDs to determinantal identities

where $x_1, \ldots, x_n$ are now independent variables, and this determinant belongs to the polynomial ring $\mathbb{Z}[x_1, \ldots, x_n]$ in n variables. We can regard Δ as a polynomial in x_i whose coefficients are polynomials in the other variables. If we evaluate this at x_j for $j \neq i$ (loosely, we put $x_i = x_j$) we obtain two columns the same, so Δ becomes zero. Thus x_j is a root of Δ regarded as a polynomial in x_i so, by the Factor Theorem (Theorem 2.8.2), $x_i - x_j$ is a factor of Δ. Thus Δ has the $\frac{1}{2}n(n-1)$ elements $x_i - x_j$ (for $1 \leq i < j \leq n$) as factors in $\mathbb{Z}[x_1, \ldots, x_n]$. Now $x_i - x_j$ (with $i < j$) is of degree 1 with leading coefficient 1 in x_i, so it is irreducible in the UFD $\mathbb{Z}[x_1, \ldots, x_n]$. It follows that the factorization of Δ into irreducible factors in $\mathbb{Z}[x_1, \ldots, x_n]$ is of the form

$$\Delta = \Phi \prod_{1 \leq i < j \leq n} (x_i - x_j),$$

where $\Phi \in \mathbb{Z}[x_1, \ldots, x_n]$. Now Δ is a homogeneous polynomial in $\mathbb{Z}[x_1, \ldots, x_n]$ of degree

$$(n-1) + (n-2) + \ldots + 1 = \tfrac{1}{2}(n-1)n,$$

and the same is true of

$$\prod_{1 \leq i < j \leq n} (x_i - x_j).$$

It follows that Φ must be homogeneous of degree zero, i.e. $\Phi \in \mathbb{Z}$. Now the product of the terms on the leading diagonal of Δ is $x_1^{n-1} x_2^{n-2} \ldots x_{n-1}$, and there is no other way in which this term can occur in the determinant. Likewise, if we look at the product

$$\prod_{1 \leq i < j \leq n} (x_i - x_j),$$

we again see that the term $x_1^{n-1} x_2^{n-2} \ldots x_{n-1}$ occurs (multiply together the first of the two terms in each bracket), and it occurs just once. It follows that $\Phi = 1$. We now apply the homomorphism $\mathbb{Z}[x_1, \ldots, x_n] \to R$ given by evaluating a polynomial $f(x_1, \ldots, x_n)$ in $\mathbb{Z}[x_1, \ldots, x_n]$ at $(a_1, \ldots, a_n)$, i.e. we map $f(x_1, \ldots, x_n)$ to $f(a_1, \ldots, a_n)$. (This is not quite straightforward, because the integer coefficient $n = 1 + \ldots + 1$ (n times) maps to $1_R + \ldots + 1_R$, $(-1) + \ldots + (-1)$ maps to $(-1_R) + \ldots + (-1_R)$ and the integer 0 maps to 0_R.) Under

this homomorphism, Δ maps to D and
$$\prod_{1 \leq i < j \leq n} (x_i - x_j) \mapsto \prod_{1 \leq i < j \leq n} (a_i - a_j).$$
Thus,
$$D = \prod_{1 \leq i < j \leq n} (a_i - a_j),$$
and we have evaluated Vandermonde's determinant.

In practice, the steps are not usually set out as painstakingly as this. In fact, writers may not introduce new variables $x_1, \ldots, x_n$ at all, but simply regard $a_1, \ldots, a_n$ as independent variables in evaluating whatever it is they are investigating. But their argument only works because $\mathbb{Z}[x_1, \ldots, x_n]$ is a UFD and an evaluation mapping $\mathbb{Z}[x_1, \ldots, x_n] \to R$ (where R is any commutative ring) is a homomorphism.

As a second illustration, we consider the *circulant*, namely the determinant
$$C = \begin{vmatrix} a_1 & a_2 & a_3 & \cdots & a_n \\ a_n & a_1 & a_2 & \cdots & a_{n-1} \\ a_{n-1} & a_n & a_1 & \cdots & a_{n-2} \\ \vdots & \vdots & \vdots & \cdots & \vdots \\ a_2 & a_3 & a_4 & \cdots & a_1 \end{vmatrix}$$
where the rows are permuted in a circular fashion and $a_1, a_2, \ldots, a_n$ are real numbers (or they could be complex numbers). Again we introduce n independent variables $x_1, \ldots, x_n$ and consider the determinant
$$\Gamma = \begin{vmatrix} x_1 & x_2 & x_3 & \cdots & x_n \\ x_n & x_1 & x_2 & \cdots & x_{n-1} \\ x_{n-1} & x_n & x_1 & \cdots & x_{n-2} \\ \vdots & \vdots & \vdots & \cdots & \vdots \\ x_2 & x_3 & x_4 & \cdots & x_1 \end{vmatrix}$$
in the polynomial ring $\mathbb{C}[x_1, \ldots, x_n]$. For each integer $k = 0, 1, \ldots, n-1$, we write
$$\omega_k = e^{2k\pi i/n},$$

2.9 Application of UFDs to determinantal identities

so that $\omega_0, \omega_1, \ldots, \omega_{n-1}$ are the distinct complex nth roots of unity. We now add to the first column of Γ

$$\omega_k(\text{column 2}) + \omega_k^2(\text{column 3}) + \ldots + \omega_k^{n-1}(\text{column } n)$$

to give the new first column

$$x_1 + \omega_k x_2 + \omega_k^2 x_3 + \ldots + \omega_k^{n-1} x_n$$
$$x_n + \omega_k x_1 + \omega_k^2 x_2 + \ldots + \omega_k^{n-1} x_{n-1}$$
$$x_{n-1} + \omega_k x_n + \omega_k^2 x_1 + \ldots + \omega_k^{n-1} x_{n-2}$$
$$\ldots\ldots\ldots\ldots\ldots\ldots\ldots\ldots\ldots\ldots\ldots\ldots\ldots\ldots\ldots$$
$$x_2 + \omega_k x_3 + \omega_k^2 x_4 + \ldots + \omega_k^{n-1} x_1$$

and these entries have the common factor

$$x_1 + \omega_k x_2 + \omega_k^2 x_3 + \ldots + \omega_k^{n-1} x_n$$

(remember that $\omega_k^n = 1$), which is thus a factor of Γ. Now this factor is irreducible in the polynomial ring $\mathbb{C}[x_1, \ldots, x_n]$ because it is of degree 1 with leading coefficient 1 in x_1. This is true for all integers k from 1 to n. Since $\mathbb{C}[x_1, \ldots, x_n]$ is a UFD, the factorization of Γ into its irreducible factors must therefore be of the form

$$\Gamma = \Phi \prod_{1 \leq k \leq n} (x_1 + \omega_k x_2 + \omega_k^2 x_3 + \ldots + \omega_k^{n-1} x_n)$$

for some $\Phi \in \mathbb{C}[x_1, \ldots, x_n]$. Now Γ is a homogeneous polynomial in $\mathbb{C}[x_1, \ldots, x_n]$ of degree n, and the same is true of the product of n linear factors on the right-hand side. It follows that $\Phi \in \mathbb{C}$. The leading diagonal of Γ gives the term x_1^n, which also occurs in the product on the right-hand side. It follows that $\Phi = 1$. We now apply the homomorphism

$$\mathbb{C}[x_1, \ldots, x_n] \to \mathbb{C}$$

under which a polynomial in $\mathbb{C}[x_1, \ldots, x_n]$ is evaluated at $(a_1, \ldots, a_n)$. This gives

$$C = \prod_{1 \leq k \leq n} (a_1 + \omega_k a_2 + \omega_k^2 a_3 + \ldots + \omega_k^{n-1} a_n).$$

Some of the exercises which follow can be solved by the more usual techniques of evaluating such determinants, namely by subtracting columns and taking factors out of columns. It is suggested alternatively that readers regard the entries as independent variables and try to spot factors directly. Careful readers will introduce new symbols for the variables in the polynomial ring and then apply an evaluation mapping to $\mathbb{R}$. Others will not bother but hopefully will have at the back of their minds that this is what they should be doing, and that the whole thing works because a polynomial ring over $\mathbb{Z}$ is a UFD.

Exercises 2.9

1. Evaluate the following determinants in which the elements belong to an arbitrary commutative ring R:

(i) $\begin{vmatrix} b^2+c^2 & c^2+a^2 & a^2+b^2 \\ b+c & c+a & a+b \\ 1 & 1 & 1 \end{vmatrix}$

(ii) $\begin{vmatrix} bc & ca & ab \\ a & b & c \\ 1 & 1 & 1 \end{vmatrix}$

(iii) $\begin{vmatrix} a_1^n & a_2^n & a_3^n & \cdots & a_n^n \\ a_1^{n-2} & a_2^{n-2} & a_3^{n-2} & \cdots & a_n^{n-2} \\ a_1^{n-3} & a_2^{n-3} & a_3^{n-3} & \cdots & a_n^{n-3} \\ \vdots & \vdots & \vdots & \cdots & \vdots \\ 1 & 1 & 1 & \cdots & 1 \end{vmatrix}$

(iv) $\begin{vmatrix} a_1^n & a_2^n & \cdots & a_n^n \\ a_1^{n-1} & a_2^{n-1} & \cdots & a_n^{n-1} \\ \vdots & \vdots & \cdots & \vdots \\ a_1^2 & a_2^2 & \cdots & a_n^2 \\ 1 & 1 & \cdots & 1 \end{vmatrix}$

(v) $\begin{vmatrix} a^3 & b^3 & c^3 & d^3 \\ a^2 & b^2 & c^2 & d^2 \\ b+c+d & a+c+d & a+b+d & a+b+c \\ 1 & 1 & 1 & 1 \end{vmatrix}$

2.9 Application of UFDs to determinantal identities

(vi) $\begin{vmatrix} bcd & acd & abd & abc \\ b^2+c^2+d^2 & a^2+c^2+d^2 & a^2+b^2+d^2 & a^2+b^2+c^2 \\ a & b & c & d \\ 1 & 1 & 1 & 1 \end{vmatrix}$

2. Evaluate the following determinant, where $a_1, \ldots, a_n$ are complex numbers:

$$\begin{vmatrix} a_1 & a_2 & a_3 & \ldots & a_n \\ -a_n & a_1 & a_2 & \ldots & a_{n-1} \\ -a_{n-1} & -a_n & a_1 & \ldots & a_{n-2} \\ \vdots & \vdots & \vdots & \ldots & \vdots \\ -a_2 & -a_3 & -a_4 & \ldots & a_1 \end{vmatrix}.$$

POSTSCRIPT

Where do we go from here? Our approach to rings has been decidedly 'old-fashioned'. We have reached the point at which mathematicians like Kummer, Dedekind and Kronecker arrived in the last century. They had come across rings such as $\mathbb{Z}[\sqrt{-5}]$ in their work in number theory, and in particular the fact that these rings may not be UFDs. The question was: how could unique factorization be restored in such rings? This was resolved in the following way. Instead of factorizing a single element, they took a set of elements of the ring with certain properties, called an 'ideal'. It was then shown by Dedekind how, in these 'rings of algebraic integers' such as $\mathbb{Z}[\sqrt{-5}]$, an ideal could be factorized uniquely as a product of 'prime ideals'. This led to the study of 'Dedekind rings' and 'Multiplicative (or Classical) Ideal Theory'.

As well as in number theory, rings were also arising in algebraic geometry, but they were different sorts of rings. Algebraic geometry is concerned with where polynomials vanish (think of a parabola as the points where the polynomial $y^2 - 4ax$ vanishes), so the rings which arise are polynomial rings in a finite number (>1) of variables over the field $\mathbb{C}$ of complex numbers (to make it easier!). These rings are not Dedekind rings. At the turn of the century, the German mathematician David Hilbert showed that the ideals of these polynomial rings are 'finitely generated'. In the first part of the 20th century, the great German mathematician Emmy Noether took Hilbert's ideas and presented them in an abstract context. She showed that, in certain rings including Dedekind rings and polynomial rings in finitely many variables over a field, every ideal is the intersection (not the product) of certain sorts of ideals called 'primary ideals', and that connected with this intersection are some prime ideals which are uniquely associated with the original ideal. So the theme of unique factorization, going right back to the ancient Greeks, is retained in the beautiful results of Emmy Noether, albeit in a much transformed way.

The author first learned about Emmy Noether's work from the pioneering tract by his teacher D. G. Northcott entitled *Ideal Theory*, and is happy to recommend this as a beautiful account of an extremely elegant piece of mathematics. Other books on ring theory, more comprehensive and advanced than the present one, also contain accounts of 'Noetherian' and 'Dedekind rings'.

Just a word in closing about more recent results concerning unique factorization. One of the most notable achievements of a branch of modern algebra called 'homological algebra' has been the proof that a class of rings which arise in algebraic geometry, called 'regular local rings' are UFDs. This was proved in different ways by various people around 1960. We have proved here that, when R is a UFD, then so is the polynomial ring $R[x]$. It may be asked whether the same is true for a power-series ring. Pierre Samuel in 1963 gave an example of a UFD R for which the power-series ring $R[[x]]$ is not a UFD. It may then be asked: for which UFDs R is $R[[x]]$ a UFD? For example, it is true that $\mathbb{Z}[[x]]$ is a UFD and, more generally, if R is a 'principal ideal domain', then $R[[x]]$ is a UFD. So there is a lot more that can be said about UFDs, and in particular about how they fit into the wider picture of the theory of commutative rings. (The book *Commutative Rings* by Irving Kaplansky, who is one of those who proved that a regular local ring is a UFD, contains a lot of information in this direction.) We hope that some of our readers might have had their interest aroused for further study. There is a lot of fascinating mathematics ahead!

PARTIAL SOLUTIONS TO EXERCISES

Exercises 1.3
1. This is a routine verification of the axioms; there is no short cut!
2. Again routine.
3. For the first result, consider $(-a)^2$; for the second, expand $(a+b)^2$. But be warned, $(a+b)^2$ expanded is not $a^2+2ab+b^2$! You should get $ab = -ba$; you want $ab = ba$.
4. Everything is routine, and somewhat tedious, except for the verification that $\triangle$ is associative. You could draw a 'Venn diagram', but that is cheating! Note that $X \backslash Y = X \cap Y'$, where Y' denotes the complement of Y in S, i.e. $Y' = \{s \in S : s \notin Y\}$. Show that $X \triangle (Y \triangle Z)$ and $(X \triangle Y) \triangle Z$ are both equal to

$$(X \cap Y' \cap Z') \cup (X' \cap Y \cap Z') \cup (X' \cap Y' \cap Z) \cup (X \cap Y \cap Z)$$

using the distributive laws for $\cap$, $\cup$ and de Morgan's Laws, namely

$$(X \cup Y)' = X' \cap Y', (X \cap Y)' = X' \cup Y'.$$

Exercises 1.4
1. It is hoped that you did not fall into the trap of checking all the axioms for a ring. Use the Subring Criterion to show that this is a subring of $\mathbb{Q}$. The property of a prime number that you use is that, if p does not divide integers n_1, n_2, then it does not divide their product $n_1 n_2$.
2. Again, use the Subring Criterion to show that this is a subring of $M_2(\mathbb{Z})$.
3. Note that $\omega^3 = 1$ so that $\omega^2 + \omega + 1 = 0$. This is the subring $\mathbb{Z}[\omega]$ of $\mathbb{C}$.
4. This is the subring $\mathbb{Z}[(\sqrt{(n)} - 1)/2]$ of $\mathbb{C}$. Note that

$$\left(\frac{\sqrt{(n)}-1}{2}\right)^2 = \frac{n-1}{4} - \frac{\sqrt{(n)}-1}{2}$$

and $(n-1)/4 \in \mathbb{Z}$ because $n \equiv 1 \pmod 4$.
5. The mapping $f: R \to R$ given by $f(a) = 1 - a$ $(a \in R)$ reverses 0 and 1.

Partial solutions to exercises 93

This is a bijection. Define $\oplus$, $\odot$ on R by

$$a \oplus b = f^{-1}(f(a)+f(b)) = a+b-1,$$
$$a \odot b = f^{-1}(f(a)f(b)) = a+b-ab.$$

Then $(R, \oplus, \odot)$ is a ring and f is an isomorphism from $(R, \oplus, \odot)$ to $(R, +, \cdot)$.

Exercises 1.5

1. Try $(1, 0)(0, 1)$, $(0, 1)(0, 1)$, $X \triangle X$. The only Boolean rings which are integral domains are those with just two elements, 0 and 1. In 1.4, three of the rings are subrings of integral domains, and so are themselves integral domains. In Exercise 1.4.2, try

$$\begin{bmatrix} 1 & 0 \\ 0 & 0 \end{bmatrix} \begin{bmatrix} 0 & 0 \\ 0 & 1 \end{bmatrix}.$$

2. (a) $\operatorname{ord}(f(x)g(x)) = \operatorname{ord} f(x) + \operatorname{ord} g(x)$. (b) $\operatorname{ord}(f(x)g(x)) \geqslant \operatorname{ord} f(x) + \operatorname{ord} g(x)$. The conventions for ∞ are $\infty + \infty = \infty$, $\infty + n = \infty$, $\infty > n$ (n a non-negative integer). The formula in (a) will show that, when R is an integral domain, so is $R[[x]]$.

Exercises 1.6

1. The Subring Criterion shows that $\mathbb{Q}[\sqrt{2}]$ is a subring of $\mathbb{R}$, and so is an integral domain. For a non-zero element $a+b\sqrt{2}$ (with $a, b \in \mathbb{Q}$),

$$(a+b\sqrt{2})^{-1} = \frac{a}{a^2-2b^2} - \frac{b}{a^2-2b^2}\sqrt{2}.$$

But why is $a^2 - 2b^2$ not zero?

2. $x\bar{x} = \bar{x}x = (a^2+b^2+c^2+d^2)I$, so $x \neq 0$ has inverse $\bar{x}/(a^2+b^2+c^2+d^2)$. As to whether $\mathbb{H}$ is a field, consider AB and BA. To embed $\mathbb{C}$ in $\mathbb{H}$, map $a+bi$ $(a, b \in \mathbb{R})$ to $aI+bA$; or $aI+bB$; or $aI+bC$.

3. If $a_0 + a_1 x + a_2 x^2 + \ldots$ is a unit, multiply it by its inverse and equate constant terms to show that a_0 is a unit. Conversely, if a_0 is a unit, then the coefficients of the inverse $b_0 + b_1 x + b_2 x^2 + \ldots$ of $a_0 + a_1 x + a_2 x^2 + \ldots$ can be defined inductively: $b_0 = a_0^{-1}$ and, assuming that $b_0, b_1, \ldots, b_{n-1}$ have been defined, define

$$b_n = -a_0^{-1}(a_1 b_{n-1} + a_2 b_{n-2} + \ldots + a_n b_0),$$

so that $a_0 b_n + a_1 b_{n-1} + \ldots + a_n b_0 = 0$.

$$(1+x)^{-1} = 1 - x + x^2 - x^3 + \ldots.$$

A non-zero power series in $k[[x]]$ is of the form $a_n x^n + a_{n+1} x^{n+1} + \ldots$, where $a_n \neq 0$, which can be written as $x^n u$, where $u = a_n + a_{n+1} x + \ldots$. Now a_n is a unit of k, so u is a unit of $k[[x]]$.

4. Use the Subring Criterion to show that F is a subring of $M_2(\mathbb{R})$, and so is a ring. Check that it is commutative. Finally,

$$\begin{bmatrix} a & b \\ -b & a \end{bmatrix}^{-1} = \begin{bmatrix} \dfrac{a}{a^2+b^2} & \dfrac{-b}{a^2+b^2} \\ \dfrac{b}{a^2+b^2} & \dfrac{a}{a^2+b^2} \end{bmatrix} \in F,$$

where a, b are not both zero.

5. Either check all the axioms for a field, which is tedious, or spot the bijection $\mathbb{R} \times \mathbb{R} \to F$ (with F as in Exercise 1.6.4) given by

$$(a, b) \mapsto \begin{bmatrix} a & b \\ -b & a \end{bmatrix}.$$

This makes $\mathbb{R} \times \mathbb{R}$ into a ring isomorphic to F, with addition and multiplication as given, so that it is a field. Note that $(0, 1)^2 = -(1, 0)$. This is the field of complex numbers with (a, b) more usually written as $a + ib$.

6. The first approach is to show that these form a subring of $M_2(\mathbb{Q})$, and so form a ring. Then show that this ring is commutative and

$$\begin{bmatrix} a & 2b \\ b & a \end{bmatrix}^{-1} = \begin{bmatrix} \dfrac{a}{a^2-2b^2} & \dfrac{2(-b)}{a^2-2b^2} \\ \dfrac{-b}{a^2-2b^2} & \dfrac{a}{a^2-2b^2} \end{bmatrix}$$

when a, b are not both zero. Why is $a^2 - 2b^2 \neq 0$? The second approach is to spot a connection with the field $\mathbb{Q}[\sqrt{2}]$ in Exercise 1.6.1. Define a mapping from this set of matrices to $\mathbb{Q}[\sqrt{2}]$ by

$$\begin{bmatrix} a & 2b \\ b & a \end{bmatrix} \mapsto a + b\sqrt{2}.$$

This is a bijection, and makes the set of matrices into a ring isomorphic to $\mathbb{Q}[\sqrt{2}]$, with matrix addition and multiplication as the binary operations. Thus we have a field.

7. There will exist non-negative integers r, s with $r < s$ and $a^r = a^s$. Then $a^r(a^{s-r} - 1) = 0$. Deduce that $a^{s-r} = 1$ and hence that a has inverse a^{s-r-1}. In $\mathbb{Z}_7$, $\bar{1}^{-1} = \bar{1}^0, \bar{2}^{-1} = \bar{2}^2, \bar{3}^{-1} = \bar{3}^5, \bar{4}^{-1} = \bar{4}^2, \bar{5}^{-1} = \bar{5}^5, \bar{6}^{-1} = \bar{6}^1$.

Exercises 1.7

1. By Fermat's Little Theorem, $3^{22} \equiv 1 \pmod{23}$. Write $3^{47} = (3^{22})^2 3^3 \equiv 1^2 \times 3^3 \equiv 4 \pmod{23}$, so the remainder is 4.

2. Try $k = 4, n = 2$.

3. Careful! 561 is not prime so Fermat's Little Theorem does not give the congruence immediately. In prime factors, $561 = 3 \times 11 \times 17$. By Fermat,

Partial solutions to exercises 95

$n^3 \equiv n \pmod 3$. Multiply by n^2 to give $n^5 \equiv n^3 \pmod 3$ so $n^5 \equiv n^3 \equiv n \pmod 3$. Then $n^7 \equiv n^5 \equiv n^3 \equiv n \pmod 3$ and so on, giving finally that $n^{561} \equiv n \pmod 3$. Similarly, $n^{11} \equiv n \pmod{11}$ so $n^{21} \equiv n^{11} \equiv n \pmod{11}$, $n^{31} \equiv n^{21} \equiv n^{11} \equiv n \pmod{11}$, and so on, giving finally $n^{561} \equiv n \pmod{11}$. Similarly $n^{1+16+\cdots+16} \equiv n \pmod{17}$ and $561 = 1 + (16 \times 35)$, so $n^{561} \equiv n \pmod{17}$. If an integer is divisible by the primes 3, 11, 17, then it is divisible by $3 \times 11 \times 17$. (See Exercise 2.4.7.) Thus $n^{561} - n$ is divisible by 561. (Positive integers k which satisfy $n^k \equiv n \pmod k$ for all $n \in \mathbb{Z}$ and yet are not prime are called *Carmichael numbers*. Other examples are 1105, 1729, 2465. It is thought that there are infinitely many such numbers.)

4. $1001 = 7 \times 11 \times 13$. $143^6 + 91^{10} + 77^{12} \equiv 1 + 0 + 0 \pmod 7$, $\equiv 0 + 1 + 0 \pmod{11}$, $\equiv 0 + 0 + 1 \pmod{13}$, so $\equiv 1 \pmod{7 \times 11 \times 13}$.

Exercise 1.7.3.

6. Reduce 2222, 5555 modulo 7. Now write the powers as $6 \times 925 + 5$, $6 \times 370 + 2$. Then $2222^{5555} + 5555^{2222} \equiv 3^5 + 4^2 \pmod 7$ etc.

7. Suppose that n is not prime. Then there exists $m \in \mathbb{Z}$ such that $1 < m < n$ and m divides n. Suppose that $(n-1)! + 1 = kn$ for some $k \in \mathbb{Z}$. Since m divides $(n-1)!$, it follows that m divides 1, which gives a contradiction.

$$16! = 16 \times 15 \times 14 \times \ldots \times 1$$
$$\equiv (-1) \times (-2) \times \ldots \times (-8) \times 8! \pmod{17}$$
$$\equiv (2 \times 8)^2 \times (3 \times 5)^2 \times (4 \times 6)^2 \times 7^2 \pmod{17}$$
$$\equiv (-1)^2 \times (-2)^2 \times 7^2 \times 7^2 \pmod{17} \text{ etc.}$$

8. If n is prime, there exists m such that $k = 0$ by Wilson, and, with this m, $p = n$. If n is not prime, k is never zero by Exercise 1.7.7, and $p = 2$. To obtain $p = 7$, choose $n = 7$ and m so that $6! + 1 = 7m$.

9. By Fermat and Wilson, $n^p + (p-1)!n \equiv n - n \pmod p$.

10. By Wilson,

$$1 \times 2 \times \ldots \times (p-1) \equiv -1 \pmod p.$$

Subtract p from every even term in the product. This will not affect the congruence modulo p, and we obtain

$$1 \times (2-p) \times 3 \times (4-p) \times \ldots \times (p-2) \times (-1) \equiv -1 \pmod p,$$
$$(-1)^{(p-1)/2}(1^2 \times 3^2 \times \ldots \times (p-2)^2) \equiv -1 \pmod p,$$

which gives the result.

Exercises 2.2

1. $6 = 3 \times 2 = (1 + \sqrt{-5})(1 - \sqrt{-5})$. $N(3) = 9$, $N(2) = 4$, $N(1 \pm \sqrt{-5}) = 6$. In Section 2.2 we showed that 3 is irreducible; it is because there are no elements of norm 3. Similarly, there is no element of norm 2. It follows by a similar technique that $2, 1 \pm \sqrt{-5}$ are irreducible. Also, none of these are associates of each other, because the only units are ± 1.

96   *Partial solutions to exercises*

2. (*a*) A straightforward verification.

(*b*) If α is a unit of $\mathbb{Z}[\sqrt{5}]$, then $\alpha\alpha^{-1}=1$. Use (*a*) to show that $N(\alpha)=\pm 1$. Conversely, suppose that $\alpha = a+b\sqrt{5}$ (a, $b\in\mathbb{Z}$) and that $N(\alpha)=\pm 1$. Then $a^2 - 5b^2 = \pm 1$, so $(a+b\sqrt{5})(a-b\sqrt{5}) = \pm 1$ and $\alpha^{-1} = \pm(a-b\sqrt{5})$. $9+4\sqrt{5}$ is a unit with norm 1 and $2+\sqrt{5}$ is a unit with norm -1.

(*c*) Let $N(\alpha)$ be prime, and write $\alpha = \beta\gamma$, where $\beta, \gamma \in \mathbb{Z}[\sqrt{5}]$. Then $N(\alpha) = N(\beta)N(\gamma)$, so either $N(\beta) = \pm 1$ or $N(\gamma) = \pm 1$. Hence, either β or γ is a unit.

(*d*) For $x\in\mathbb{Z}$, $x \equiv 0, 1, 2, 3$ or 4 (mod 5), so $x^2 \equiv 0^2, 1^2, 2^2, 3^2$ or 4^2 (mod 5), i.e. $x^2 \equiv 0, 1$ or 4 (mod 5). Now 2, $3\pm\sqrt{5}$ have norm 4. Consider $\alpha\in\mathbb{Z}[\sqrt{5}]$ of norm 4, and write $\alpha = \beta\gamma$, where $\beta, \gamma \in \mathbb{Z}[\sqrt{5}]$. Then $N(\beta)$, $N(\gamma)$ divide 4 in $\mathbb{Z}$. Put $\beta = a+b\sqrt{5}$, where $a, b\in\mathbb{Z}$. If $N(\beta) = \pm 2$, then $a^2 - 5b^2 = \pm 2$ and $a^2 \equiv \pm 2$ (mod 5), which is impossible. Thus $N(\beta) = \pm 1$ or $N(\gamma) = \pm 1$, so β or γ is a unit and α is irreducible.

(*e*) You need to point out that $2 \nmid 3+\sqrt{5}$. But, if $2|3+\sqrt{5}$ in $\mathbb{Z}[\sqrt{5}]$, then $2|1$ in $\mathbb{Z}$.

Exercises 2.3

1. $\dfrac{11+7i}{2+5i} = \dfrac{(11+7i)(2-5i)}{29} = \dfrac{57-41i}{29}.$

The 'nearest' Gaussian integer is $2-i$, so write

$$11+7i = (2-i)(2+5i) + 2-i.$$

We can take $2-i$ as quotient and $2-i$ as remainder. If we do not choose the 'nearest' Gaussian integer, we could for example try $2-2i$ as quotient. This gives

$$11+7i = (2-2i)(2+5i) - 3+i.$$

It is true that $|-3+i| < |2+5i|$. (This is no longer automatically so, because we have not taken the nearest Gaussian integer.) Thus we can take $2-2i$ as quotient and $-3+i$ as remainder, so quotient and remainder are not unique.

2. Make minor alterations to the proof of Theorem 2.3.4. Note that, for $\lambda, \mu \in \mathbb{Z}$,

$$N(\lambda + \mu\sqrt{-2}) = \lambda^2 + 2\mu^2.$$

3. Obvious from property 1 of Definition 2.3.3.

4. Obvious from property 1 of Definition 2.3.3, Exercise 2.3.3 and Theorem 2.3.5.

5. (*a*) $\omega = (1+i\sqrt{3})/2$ and $\omega^3 = -1$, so that $1-\omega+\omega^2 = 0$ and $\omega^2 = \omega - 1$. Thus a typical element of R is of the form $a = m + n\omega$, where

Partial solutions to exercises 97

$m, n \in \mathbb{Z}$, and, if $a \neq 0$,

$$\partial(a) = \left(m + \frac{n}{2}\right)^2 + \frac{3n^2}{4} = m^2 + mn + n^2,$$

a positive integer. Let $a, b \in R \setminus \{0\}$ with $a | b$. Then there exists $c \in R \setminus \{0\}$ such that $b = ac$ and $\partial(b) = |b|^2 = |a|^2 |c|^2 = \partial(a) \partial(c) \geq \partial(a)$. Now consider $a, b \in R$ with $b \neq 0$. Write $a = m + n\omega$, $b = r + s\omega$ ($m, n, r, s \in \mathbb{Z}$). Now $\bar{\omega} = e^{-\pi i/3} = \omega^{-1} = -\omega^2 = 1 - \omega$, and

$$\frac{a}{b} = \frac{m + n\omega}{r + s\omega} = \frac{(m + n\omega)(r + s\bar{\omega})}{(r + s\omega)(r + s\bar{\omega})}$$

$$= \frac{mr + ns + nr\omega + ms\bar{\omega}}{r^2 + s^2 + rs(\omega + \bar{\omega})}$$

$$= \frac{mr + ns + ms + (nr - ms)\omega}{r^2 + s^2 + rs}$$

$$= \lambda + \mu\omega$$

for some $\lambda, \mu \in \mathbb{Q}$. We can write $\lambda = \lambda_1 + \lambda_2$, $\mu = \mu_1 + \mu_2$, where $\lambda_1, \mu_1 \in \mathbb{Z}$, $\lambda_2, \mu_2 \in \mathbb{Q}$, $|\lambda_2| \leq \frac{1}{2}$, $|\mu_2| \leq \frac{1}{2}$. Thus

$$a = b(\lambda_1 + \mu_1 \omega) + b(\lambda_2 + \mu_2 \omega).$$

Put $q = \lambda_1 + \mu_1 \omega$, $r = b(\lambda_2 + \mu_2 \omega)$. Then $q \in R$ so $r = a - bq \in R$, and, if $r \neq 0$,

$$\partial(r) = |b|^2 |\lambda_2 + \mu_2 \omega|^2 = \partial(b)(\lambda_2^2 + \lambda_2 \mu_2 + \mu_2^2)$$
$$\leq \tfrac{3}{4}\partial(b) < \partial(b).$$

This shows that R is a Euclidean domain.

(b) Let $\partial(a) = 1$. Write $a = m + n\omega$ with $m, n \in \mathbb{Z}$. Then

$$\left(m + \frac{n}{2}\right)^2 + \frac{3n^2}{4} = 1.$$

If $n = 0$, then $m = \pm 1$. If $n = 1$ then $m = 0$ or -1. If $n = -1$ then $m = 0$ or 1. This gives the elements ± 1, ω, $-1 + \omega$, $-\omega$, $1 - \omega$ or ± 1, $\pm \omega$, $\pm \omega^2$. These elements do all have degree 1, and so are precisely the elements of degree 1.

(c) A unit must have degree 1 and so must be one of ± 1, $\pm \omega$, $\pm \omega^2$. Since these are all units, they are therefore all the units.

(d) Write $i\sqrt{3} = ab$ with $a, b \in R$. Then $3 = \partial(a)\partial(b)$ so either $\partial(a) = 1$ or $\partial(b) = 1$. Thus either a or b is a unit. Hence $i\sqrt{3}$ is irreducible.

(e) Suppose that $i\sqrt{3} | (m + n\omega)$, where $m, n \in \mathbb{Z}$. Then $3 | \partial(m + n\omega)$ in $\mathbb{Z}$, so $3 | (m^2 + mn + n^2)$ so $3 | (m - n)^2 + 3mn$ so $3 | (m - n)^2$ so $3 | (m - n)$ in $\mathbb{Z}$.

Conversely, write $m-n=3r$, where $r\in\mathbb{Z}$. Then

$$m+n\omega = n+3r+\tfrac{1}{2}n(1+i\sqrt{3})$$
$$= i\sqrt{3}\left(-i\sqrt{3}r+n\frac{1-i\sqrt{3}}{2}\right)$$
$$= i\sqrt{3}((1-2\omega)r+n(1-\omega)),$$

and $i\sqrt{3}\mid m+n\omega$ in R.

(*f*) Suppose that $i\sqrt{3}\nmid a$ in R, and write $a=m+n\omega$, where $m, n\in\mathbb{Z}$. Then $3\nmid m-n$ in $\mathbb{Z}$. There are three possibilities:
(1) $m=3k, n=3l+t$ for some $k, l\in\mathbb{Z}$ and $t=\pm 1$;
(2) $m=3k+t, n=3l$ for some $k, l\in\mathbb{Z}$ and $t=\pm 1$;
(3) $m=3k+t, n=3l-t$ for some $k, l\in\mathbb{Z}$ and $t=\pm 1$.
In case (1),

$$a = 3b+t\omega, \text{ where } b=k+l\omega\in R,$$

so

$$a^3 = 27b^3 + 27b^2 t\omega + 9bt^2\omega^2 + t^3\omega^3 \equiv \pm 1 \pmod{9}.$$

Cases (2), (3) are similar.

Exercises 2.4

1. Let γ be a common divisor of $3, 2+\sqrt{-5}$ in $\mathbb{Z}[\sqrt{-5}]$. Then $3=\gamma\delta$ and $2+\sqrt{-5}=\gamma\tau$ for some $\delta, \tau\in\mathbb{Z}[\sqrt{-5}]$. Take norms as in Section 2.2 and show that $\gamma=\pm 1$. Deduce that 1 is GCD of $3, 2+\sqrt{-5}$. Try writing

$$1 = 3(a+b\sqrt{-5}) + (2+\sqrt{-5})(c+d\sqrt{-5}),$$

where $a, b, c, d\in\mathbb{Z}$. Equate real and imaginary parts and deduce that a, b, c, d cannot exist.

2. For the first part, use induction on the number of elements. It is obvious for one element. Now consider $a_1, \ldots, a_n\in R$ with $n>1$, and assume that $a_1, \ldots, a_{n-1}$ have GCD c. Put $d=\text{GCD}(c, a_n)$. Show that d is GCD of $a_1, \ldots, a_n$. The formula

$$\text{GCD}(a_1, \ldots, a_n) = \text{GCD}(\text{GCD}(a_1, \ldots, a_{n-1}), a_n)$$

thus established will enable the second statement to be proved, again by induction on n.

3. Use Euclid's Algorithm as in Examples 2.4.3 and 2.4.4. The GCD is $1-\sqrt{-2}$ and you can take $\alpha = -1-2\sqrt{-2}$, $\beta = -2+\sqrt{-2}$.

4. Use Euclid's Algorithm. Do not forget that you are working with coefficients in $\mathbb{Z}_5$. You should avoid any fractions. Thus $\bar{2}^{-1} = \bar{3}$, $\bar{3}^{-1} = \bar{2}$, $\bar{4}^{-1} = \bar{4}$. When you divide $g(x)$ by $f(x)$ you should obtain the remainder $\bar{3}x^4 + x^3 + \bar{2}x^2 + \bar{2}x + \bar{2}$. Now divide $f(x)$ by this polynomial to give the remainder $\bar{4}x^2 + \bar{3}x + \bar{3}$. The next division gives zero remainder, so

Partial solutions to exercises 99

$4x^2+\bar{3}x+\bar{3}$ is a GCD. To make it monic, multiply by $\bar{4}$, a unit, to give $x^2+\bar{2}x+\bar{2}$. The successive divisions give

$$s(x)=\bar{3}x^2+\bar{2}x+\bar{3},\ t(x)=\bar{2}x+\bar{1},$$

although there are other possibilities for these.

  5. Use Euclid's Algorithm to find the GCD of $4+9i$, $2+7i$ in $\mathbb{Z}[i]$. The division of $4+9i$ by $2+7i$ gives quotient 1 and remainder $2+2i$; the division of $2+7i$ by $2+2i$ gives quotient $2+i$ and remainder i. Thus the GCD is a unit, and you obtain

$$i=-(2+i)(4+9i)+(3+i)(2+7i)$$

from the two divisions. Thus

$$3+i=(1-3i)i=(-5+5i)(4+9i)+(6-8i)(2+7i).$$

  6. Show that, if $c|a$ and $c|b$, then $c|1$. Deduce that 1 is GCD of a, b.
  7. There exist $s,t\in R$ such that $sa+tb=1$. Thus $sac+tbc=c$. Show that ab divides the left-hand side of this equation.

Exercises 2.5
  1. You should find such an element if you look at Exercise 2.2.2.
  2. The fact that the binomial coefficient is an integer follows by induction on p from the identity

$$\binom{p-1}{k-1}+\binom{p-1}{k}=\binom{p}{k},$$

where $p>k\geqslant 1$. (Or you could point out that the binomial coefficient gives the number of subsets containing precisely k elements of a set of p elements.) Now let p be prime and $1\leqslant k\leqslant p-1$. Then the product

$$k!(p-k)!\binom{p}{k}$$

is divisible by p, so one of the factors must be divisible by p.

Exercises 2.7
  1. In prime factors,

$$\begin{aligned}3965&=5\times 13\times 61\\&=(2^2+1^2)(3^2+2^2)(6^2+5^2)\\&=(2+i)(2-i)(3+2i)(3-2i)(6+5i)(6-5i)\\&=(2+i)(3+2i)(6+5i)(2-i)(3-2i)(6-5i)\\&=(-11+62i)(-11-62i)\\&=11^2+62^2.\end{aligned}$$

Also,

$$3965 = |(2+i)(3+2i)(6-5i)|^2 = 59^2 + 22^2,$$
$$= |(2+i)(3-2i)(6+5i)|^2 = 53^2 + 34^2,$$
$$= |(2-i)(3+2i)(6+5i)|^2 = 43^2 + 46^2.$$

2. In prime factors, $4430 = 2 \times 5 \times 443$, and $443 \equiv 3 \pmod 4$. Look at Theorem 2.7.6. Also,

$$4430 = (1^2 + 1^2)(2^2 + 1^2)443$$
$$= (1+i)(1-i)(2+i)(2-i)443.$$

Now look at Theorem 2.7.3. Next,

$$4437 = 3^2 \times 17 \times 29$$
$$= 3^2(4^2 + 1^2)(5^2 + 2^2)$$
$$= 3^2(4+i)(4-i)(5+2i)(5-2i).$$

Now look at Theorem 2.7.3. Finally,

$$4437 = 3^2(4+i)(5+2i)(4-i)(5-2i)$$
$$= 3^2|18+13i|^2$$
$$= 54^2 + 39^2.$$

3. A factorization of a Gaussian integer will give a factorization of its norm, so we first factorize the norms into prime factors.

(i) $(3+i)(3-i) = 10 = 2 \times 5 = (1^2+1^2)(2^2+1^2)$
$$= (1+i)(1-i)(2+i)(2-i),$$

and these factors are irreducible. We now look for two factors whose product is $3+i$, namely $(1+i)(2-i)$.

(ii) $(4+3i)(4-3i) = 25 = 5^2 = ((2+i)(2-i))^2$, and $2 \pm i$ are irreducible. In this case, we cannot choose two of these factors to give $4+3i$. But $(2-i)^2 = 3-4i$ and

$$4+3i = i(3-4i) = (2-i)(1+2i),$$

and $2-i$, $1+2i$ are irreducible.

(iii) $(75+28i)(75-28i)$
$$= 6409$$
$$= 13 \times 17 \times 29$$
$$= (3^2+2^2)(4^2+1^2)(5^2+2^2)$$
$$= (3+2i)(3-2i)(4+i)(4-i)(5+2i)(5-2i)$$

in irreducibles. Now

$$(3-2i)(4-i)(5-2i) = 28 - 75i,$$

Partial solutions to exercises

so
$$75 + 28i = i(28 - 75i) = (2 + 3i)(4 - i)(5 - 2i)$$

in irreducible factors.

4. The proof of Theorem 2.7.2 will give that the primes you want are those not congruent to 3 (mod 4).

5. Use the Subring Criterion to show that F is a subring of $M_2(\mathbb{Z}_p)$. Check directly that two elements of F commute. Suppose that $p \equiv 3 \pmod 4$, and consider a non-zero element

$$\begin{bmatrix} \bar{a} & \bar{b} \\ -\bar{b} & \bar{a} \end{bmatrix}$$

of F. This has determinant $\bar{a}^2 + \bar{b}^2$. Suppose that $\bar{a}^2 + \bar{b}^2 = \bar{0}$. Then $\overline{a^2 + b^2} = \bar{0}$, so that $p|(a^2 + b^2)$ in $\mathbb{Z}$ and so also in $\mathbb{Z}[i]$. Thus $p|(a+ib)(a-ib)$ in $\mathbb{Z}[i]$ and p is irreducible in $\mathbb{Z}[i]$ (Theorem 2.7.2) so either $p|(a+ib)$ or $p|(a-ib)$ in $\mathbb{Z}[i]$. Deduce that $p|a$ and $p|b$ in $\mathbb{Z}$, so that $\bar{a} = \bar{b} = \bar{0}$, which is not true. Hence, $\bar{a}^2 + \bar{b}^2 \neq \bar{0}$ and the matrix has inverse

$$\frac{1}{\bar{a}^2 + \bar{b}^2} \begin{bmatrix} \bar{a} & -\bar{b} \\ \bar{b} & \bar{a} \end{bmatrix},$$

which belongs to F. Thus F is a field. Now suppose that $p \not\equiv 3 \pmod 4$. Then $p = a^2 + b^2$ for some $a, b \in \mathbb{Z}$ (Theorem 2.7.2), so $\bar{a}^2 + \bar{b}^2 = \bar{0}$. Show that $\bar{a} \neq \bar{0}$, $\bar{b} \neq \bar{0}$. Thus the non-zero element

$$\begin{bmatrix} \bar{a} & \bar{b} \\ -\bar{b} & \bar{a} \end{bmatrix}$$

of F has zero determinant, and so does not have an inverse. Hence, F is not a field.

There are p choices for each of $\bar{a}$, $\bar{b}$, so F possesses p^2 elements.

The problem of the Yellow Brick Road

I recall that the *Yorkshire Post* reported that only one correct solution was received from readers. (Was it from the person who set the problem?!) I have to admit that I do not have a complete solution.

Denote the distance between successive stations on the railway by d, in

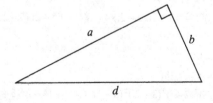

miles, and consider a right-angled triangle of roads and rail as shown. If d is even, then, because $d^2 = a^2 + b^2$, it is easy to see that a, b must also be even. Thus all distances can be divided by 2, which contradicts the fact that they are as short as possible. Thus d is odd. Suppose that d has a prime factor $p \equiv 3 \pmod{4}$. Now $p|(a^2+b^2)$ and p is irreducible in $\mathbb{Z}[i]$, so either $p|(a+ib)$ or $p|(a-ib)$ in $\mathbb{Z}[i]$. Thus $p|a$ and $p|b$ in $\mathbb{Z}$, and again distances are not as short as possible. Thus every prime factor of d is congruent to 1 (mod 4). The problem now amounts to finding the smallest positive integer d whose square can be expressed as the sum of two squares in 40 different ways.

Suppose that d is prime. We can write $d = x^2 + y^2$, with x, y positive integers, in essentially only one way (Theorem 2.7.4). Consider $d^2 = a^2 + b^2$, with a, b positive integers. Then, with $z = x + iy$,

$$z\bar{z}z\bar{z} = d^2 = (a+ib)(a-ib)$$

and $z, \bar{z}$ are irreducible in $\mathbb{Z}[i]$. By unique factorization, the only possibilities for $a+ib$ are $z^2, \bar{z}^2, z\bar{z}$ or units of $\mathbb{Z}[i]$ times these. The last of these is no good, since then a or b is zero. In the other cases, a, b are $|x^2 - y^2|, 2xy$. Thus, when d is prime, only one expression is possible.

When d is a product of two primes say $d = d_1 d_2$ then, as above, we can write $d_1 = z_1 \bar{z}_1$, $d_2 = z_2 \bar{z}_2$, and

$$z_1 z_1 z_2 z_2 \bar{z}_1 \bar{z}_1 \bar{z}_2 \bar{z}_2 = d^2 = (a+ib)(a-ib).$$

This time there are 3^2 possibilities for $a+ib$:

$$z_1 z_1 z_2 z_2, \quad z_1 \bar{z}_1 z_2 z_2, \quad \bar{z}_1 \bar{z}_1 z_2 z_2,$$
$$z_1 z_1 z_2 \bar{z}_2, \quad z_1 \bar{z}_1 z_2 \bar{z}_2, \quad \bar{z}_1 \bar{z}_1 z_2 \bar{z}_2,$$
$$z_1 z_1 \bar{z}_2 \bar{z}_2, \quad z_1 \bar{z}_1 \bar{z}_2 \bar{z}_2, \quad \bar{z}_1 \bar{z}_1 \bar{z}_2 \bar{z}_2,$$

or units times these. Of these, $z_1 \bar{z}_1 z_2 \bar{z}_2$ gives one of a, b zero, and the other eight fall into conjugate pairs. This gives $(3^2 - 1)/2 = 4$ possible expressions for d when $z_1 \neq z_2$. When $z_1 = z_2$ this reduces to 2.

Analogously, when d is a product of three distinct primes we obtain $(3^3 - 1)/2 = 13$ expressions. (This reduces to 7 when two of the primes are the same and 3 when they are all the same.) When d is a product of four distinct primes, we obtain $(3^4 - 1)/2 = 40$ expressions. This is presumably where the figure 40 in the question comes from. The first four primes congruent to 1(mod 4) are 5, 13, 17, 29, so we take

$$d = 5 \times 13 \times 17 \times 29$$
$$= (2+i)(2-i)(3+2i)(3-2i)(4+i)(4-i)(5+2i)(5-2i).$$

Thus

$$d^2 = a^2 + b^2$$
$$= (a+ib)(a-ib)$$
$$= (2+i)^2(2-i)^2(3+2i)^2(3-2i)^2(4+i)^2(4-i)^2(5+2i)^2(5-2i)^2.$$

Partial solutions to exercises

One possibility for $a+ib$ is now

$$u(2+i)^2(3+2i)^2(4+i)^2(5+2i)^2 = u(-31323-6764i),$$

where u is a unit. Choose $u=-1$ to make a, b positive. Thus $a=31323$, $b=6764$. A second possibility for $a+ib$ is

$$u(2+i)^2(3+2i)^2(4+i)^2(5+2i)(5-2i) = u(-27347+16704i).$$

Choose $u=-i$ to give $a=16704$, $b=27347$. In this way all 40 possibilities can be calculated. If these are now all added up, this gives the length of the Yellow Brick Road as 1635668 miles (at least that is what I obtained!). As to whether this is the shortest distance, I cannot say. I leave it to readers to decide!

Exercises 2.8

1. If you decide to solve (a) first, you want the complex roots of x^4+15; these are the four complex values of $(-15)^{1/4}$, namely

$$(15)^{1/4}\left(\cos\frac{\pi+2k\pi}{4} + i\sin\frac{\pi+2k\pi}{4}\right),$$

where $(15)^{1/4}$ denotes the positive real fourth root of 15 and $k=0, 1, 2, 3$. This gives

$$x^4+15 = \left(x-(15)^{1/4}\left(\frac{1+i}{\sqrt{2}}\right)\right)\left(x-(15)^{1/4}\left(\frac{-1+i}{\sqrt{2}}\right)\right)$$
$$\times\left(x-(15)^{1/4}\left(\frac{-1-i}{\sqrt{2}}\right)\right)\left(x-(15)^{1/4}\left(\frac{1-i}{\sqrt{2}}\right)\right),$$

and these are the irreducible factors in $\mathbb{C}[x]$. If we fit together conjugate pairs, we obtain

$$x^4+15 = \left(\left(x-\frac{(15)^{1/4}}{\sqrt{2}}\right)^2 + \frac{\sqrt{15}}{2}\right)\left(\left(x+\frac{(15)^{1/4}}{\sqrt{2}}\right)^2 + \frac{\sqrt{15}}{2}\right)$$
$$= (x^2 - (60)^{1/4}x + \sqrt{15})(x^2 + (60)^{1/4}x + \sqrt{15}),$$

and these are the irreducible factors in $\mathbb{R}[x]$.

A neat solution is to solve (b) before (a):

$$x^4+15 = (x^2+\sqrt{15})^2 - 2\sqrt{15}x^2$$
$$= (x^2+(60)^{1/4}x+\sqrt{15})(x^2-(60)^{1/4}x+\sqrt{15}),$$

which gives (b), and these quadratic factors can then be factorized into linear factors in $\mathbb{C}[x]$ to give (a).

Any proper factorization in $\mathbb{Q}[x]$ will also be a proper factorization in $\mathbb{R}[x]$, and the only such factorization involves irrational coefficients, so the polynomial is irreducible in $\mathbb{Q}[x]$. Alternatively, use Eisenstein's Irreducibility Criterion with $p=3$ or 5.

2. (i) Try -2 as a root.

(ii) By Theorem 2.8.13, the only possible roots of the polynomial in $\mathbb{Q}$ are $\pm 1, \pm 5, \pm 25$, and it is easily verified that none of these is a root. Hence, the polynomial has no linear factor in $\mathbb{Q}[x]$, and so is irreducible in $\mathbb{Q}[x]$, and so also in $\mathbb{Z}[x]$, because it is also primitive.

(iii) This is not irreducible in $\mathbb{Z}[x]$, because it has proper factor 3. You can use Eisenstein with $p=2$ to show that it is irreducible in $\mathbb{Q}[x]$.

(iv) Try -1 as a root.

(v) Use Eisenstein (the second version) with $q=3$.

(vi) Put $x=y+a$, where a is some integer. This gives $y^5 + \ldots + (a^5 + 5a^2 + 4)$ with the coefficients of y, y^2, y^3, y^4 divisible by 5. Choose $a=1$ and use Eisenstein with $p=5$.

(vii) Put $x=y+a$, where a is some integer, to give $3y^7 + \ldots + (3a^7 + 7a^5 + 14a^3 + 7a + 16)$, with the coefficients of $y, \ldots, y^6$ all divisible by 7. Modulo 7, the constant term is congruent to $3a^7 + 2 \equiv 3a + 2$ by Fermat's Little Theorem. Choose $a=-3$ to make the constant term divisible by 7. You will now need to check that the constant term is not divisible by 49. Now apply Eisenstein with $p=7$.

(viii) Use Eisenstein (the second version) with $q=3$ (not $q=6$, which is not prime).

(ix) Put $x=y+a$ and try to choose a so that you can apply Eisenstein with $p=5$.

3. $x^3 = x \times x \times x$.

$$x^3 + \bar{1} = (x+\bar{1})(x^2 + x + \bar{1}) \text{ and } x^2 + x + \bar{1}$$

is irreducible, since $\bar{0}, \bar{1}$ are not roots. (Note that x stands for $\bar{1}x$ etc.)

$$x^3 + x = x(x^2 + \bar{1}) = x(x+\bar{1})^2.$$

$x^3 + x + \bar{1}$ is irreducible, since $\bar{0}, \bar{1}$ are not roots.

$$x^3 + x^2 = x \times x \times (x+\bar{1}).$$

$x^3 + x^2 + \bar{1}$ is irreducible.

$$x^3 + x^2 + x = x(x^2 + x + \bar{1}).$$
$$x^3 + x^2 + x + \bar{1} = (x+\bar{1})(x^2 + \bar{1}) = (x+\bar{1})^3.$$

4. Write $f(x) = g(x)h(x)$ with $g(x), h(x) \in \mathbb{Z}[x]$. Because $f(x)$ is monic, the leading coefficients of $g(x), h(x)$ are either both 1 or both -1. Now $\bar{f}(x) = \bar{g}(x)\bar{h}(x)$ and $\deg \bar{g}(x) = \deg g(x)$, $\deg \bar{h}(x) = \deg h(x)$. Suppose that $\bar{f}(x)$ is irreducible in $\mathbb{Z}_p[x]$. Then either $\deg \bar{g}(x) = 0$ or $\deg \bar{h}(x) = 0$, i.e. either $\deg g(x) = 0$ or $\deg h(x) = 0$. Thus either $g(x) = \pm 1$ or $h(x) = \pm 1$. Thus $f(x)$ is irreducible in $\mathbb{Z}[x]$. The converse is false, e.g. $x^2 + p$ is irreducible in $\mathbb{Z}[x]$ but $x^2 + \bar{p} = x^2$ is not irreducible in $\mathbb{Z}_p[x]$.

Put $f(x) = x^3 + ax^2 + bx + c$ and take $p=2$. Then $\bar{f}(x) = x^3 + \bar{a}x^2 + \bar{b}x + \bar{c}$ in $\mathbb{Z}_2[x]$. Now $\bar{f}(\bar{0}) = \bar{c} \neq \bar{0}$ because c is odd. Also $\bar{f}(\bar{1}) =$

$\bar{1}+\bar{a}+\bar{b}+\bar{c}=\overline{1+a+b+c}=\bar{1}$ because $a+b+c$ is even. Thus $\bar{0}$, $\bar{1}$ are not roots of $\bar{f}(x)$, so $\bar{f}(x)$ has no roots, and so has no linear factors. Hence $\bar{f}(x)$ is irreducible in $\mathbb{Z}_2[x]$. Hence, $f(x)$ is irreducible in $\mathbb{Z}[x]$.

5. Theorem 2.8.12 tells us that both of these rings are UFDs. Show that 2 and x in $\mathbb{Z}[x]$ have GCD 1, but that it is impossible to express 1 in the form $2s(x)+xt(x)$, where $s(x), t(x)\in\mathbb{Z}[x]$. Use Theorem 2.4.2 to deduce that $\mathbb{Z}[x]$ is not a Euclidean domain. Show that 1 is GCD of x, y in $k[x, y]$, and use a similar method to show that $k[x, y]$ is not a Euclidean domain.

6. Follow one of the methods given in Example 2.8.16. You might like to try both methods, but be warned, Kronecker's method might involve you in a lot of attempts before you come across a factor. You should obtain

$$3x^5+4x^4+6x^3-3x-2=(x^2+x+2)(3x^3+x^2-x-1)$$

in irreducible factors in $\mathbb{Z}[x]$.

Exercises 2.9

1. (i) Let x, y, z be independent variables and consider the determinant

$$\Delta = \begin{vmatrix} y^2+z^2 & z^2+x^2 & x^2+y^2 \\ y+z & z+x & x+y \\ 1 & 1 & 1 \end{vmatrix}$$

in the polynomial ring $\mathbb{Z}[x, y, z]$. Regarding this as a polynomial in x, with coefficients in $\mathbb{Z}[y, z]$, we see that, if we evaluate at y (i.e. if we put $x=y$) columns 1 and 2 are the same and we get zero. Thus $x-y$ is an irreducible factor of Δ, as are $x-z$, $y-z$. Thus the factorization of Δ into irreducibles in $\mathbb{Z}[x, y, z]$ is of the form

$$\Delta = \Phi(x-y)(x-z)(y-z),$$

where $\Phi\in\mathbb{Z}[x, y, z]$. Consider degrees to show that $\Phi\in\mathbb{Z}$, and compare coefficients of x^2y (say) to show that $\Phi=1$. Now apply a homomorphism $\mathbb{Z}[x, y, z]\to R$ to give the value of the given determinant.

(ii) $(a-b)(a-c)(b-c)$.

(iii) Using more informal language, we regard $a_1,\ldots,a_n$ as independent variables over $\mathbb{Z}$ and show that the determinant is of the form

$$\Phi \prod_{1\leq i<j\leq n}(a_i-a_j).$$

Consideration of degrees gives that Φ is homogeneous of degree 1 in $a_1, a_2,\ldots,a_n$, say

$$\Phi=m_1a_1+m_2a_2+\ldots+m_na_n$$

with $m_1,\ldots,m_n\in\mathbb{Z}$. If we interchange two of the a_i's in the determinant, this interchanges two columns and so changes the sign. Show that it also changes

the sign of
$$\prod_{1 \leq i < j \leq n} (a_i - a_j).$$
Thus Φ is unaltered by interchange of two a_i's, i.e. Φ is symmetrical in $a_1, a_2, \ldots, a_n$. Deduce that $m_1 = \ldots = m_n$. Now compare the coefficients of $a_1^n a_2^{n-2} a_3^{n-3} \ldots 1$ to deduce that $\Phi = a_1 + \ldots + a_n$. It remains to apply an evaluation mapping from $\mathbb{Z}[x_1, \ldots, x_n]$ to $\mathbb{R}$.

(iv) Regarding the a_i's as independent variables over $\mathbb{Z}$, show that the determinant is of the form
$$\Phi \prod_{1 \leq i < j \leq n} (a_i - a_j),$$
where Φ is a homogeneous polynomial of degree $n-1$ in $a_1, \ldots, a_n$ and symmetrical in $a_1, \ldots, a_n$. No a_i can occur in Φ to degree greater than 1, so
$$\Phi = m(a_1 a_2 \ldots a_{n-1} + a_1 a_2 \ldots a_{n-2} a_n + \ldots + a_2 a_3 \ldots a_n)$$
for some $m \in \mathbb{Z}$. Look at the coefficient of $a_1^n a_2^{n-1} \ldots a_{n-1}^2$ to deduce that $m = 1$. We now evaluate to give the determinant.

(v) Regard a, b, c, d as independent variables over $\mathbb{Z}$, show that the determinant is of the form
$$\Phi(a-b)(a-c)(a-d)(b-c)(b-d)(c-d)$$
and that $\Phi = -1$. This is the 'official solution'. An alternative is to write the third row as $s-a, s-b, s-c, s-d$, where $s = a+b+c+d$, split the determinant into two using this row and use Vandermonde's determinant together with elementary properties of determinants. A similar method could be used in (i). Why not go back and try it?

(vi) $(a-b)(a-c)(a-d)(b-c)(b-d)(c-d)$.

2. Follow the same argument as for the circulant, but use the complex nth roots of -1, namely $\omega_k = e^{(2k+1)\pi i/n}$ for $k = 0, 1, \ldots, n-1$.

REFERENCES

The following books and articles are referred to in the text.

Devlin, K. (1983/4) *Mathematical Spectrum* 16, 65–7

Dörrie, H. (1965) *100 Great Problems of Elementary Mathematics*, pp. 96–104. Dover

Hardy, G. H. and Wright, E. M. (1960) *An Introduction to the Theory of Numbers*, 4th edn, Chapter 14. Oxford University Press

Kaplansky, I. (1974) *Commutative Rings (Revised Edition)*. University of Chicago Press

Lenstra, A. K., Lenstra, H. W. and Lovász, L. (1982) *Mathematische Annalen* 261, 515–34

Northcott, D. G. (1953) *Ideal Theory*. Cambridge University Press

Stewart, I. and Tall, D. (1983) *Complex Analysis. The Hitch Hiker's Guide to the Complex Plane*. Cambridge University Press

INDEX

algebraic geometry 91
algebraic integers 44
associate elements 36
associative law 9

Bézout domain 47
binary operation 5
binomial coefficient 55
Boole, George 12
Boolean ring 12

cancellation law 20
Carmichael number 95
cartesian product 5
circulant 86
Classical Ideal
 Theory 90
common divisor 47
common factor 47
commutative law 9
commutative ring 11
congruence 28
content of a
 polynomial 68
coprime 51
cyclotomic
 polynomial 77

Dedekind 3, 90
Dedekind ring 90, 91
degree 43
degree function 43
degree of a
 polynomial 21

determinantal
 identity 82
Devlin, Keith 4, 34
Dirichlet 3
distributive law 9
division algorithm 43
 for integers 40
 for polynomials 40
divisor 36
Dörrie, Heinrich 46

Eisenstein 75
Eisenstein's Irreducibility
 Criterion 75
embedding 18
Euclidean domain 43
Euclid's algorithm 48
Euclid's Elements 39
evaluation 14

factor 36
Factor Theorem 63
Fermat, Pierre 31
Fermat's Last
 'Theorem' 2, 46
Fermat's Little
 Theorem 31
field 24
field of fractions 26
formal power series 7
Fraenkel, A. A. 5
Fundamental Theorem of
 Algebra 1, 64
Fundamental Theorem of
 Arithmetic 1

109

Galois, Evariste 31
Galois field 31
Gauss 2, 31
Gaussian integer 18, 43
Gauss's Lemma 68
greatest common divisor (GCD) 47
greatest common divisor (GCD) domain 47
group of units 23

Hamilton, Sir William 27
Hardy, G. H. 45
highest common factor (HCF) 47
Hilbert, David 90
homogeneous polynomial 83
homological algebra 91
homomorphism 13

ideal 3, 90
identity element 9
image of a mapping 17
indeterminate 7, 11
integral domain 20
irreducible element 37
isomorphism 18

Kaplansky, Irving 91
Kronecker 78, 90
Kronecker's method 78, 80
Kummer 3, 90

Lagrange's Interpolation Formula 78
Lagrange's Theorem 32
Leibniz 33
Lenstra, A. K. 81
Lenstra, H. W. 4, 81
linear factor 65
Lovász, L. 81

monic polynomial 35
Multiplicative Ideal Theory 90

negative 9
Noether, Emmy 90
Noetherian ring 91
norm 38, 43
Northcott, D. G. 91

order of a power series 22

polynomial 7
polynomial ring 11
power series 7
prime element 53
prime ideal 90
prime number 37, 53
primitive polynomial 68
principal ideal domain 91
proper divisors of zero 20
proper factor 36
Public-Key Cryptography 3

quadratic polynomial 66
quaternion 27
quotient 40

regular local ring 91
relatively prime 51
remainder 40
Remainder Theorem 63
residue class 29
ring 9

Samuel, Pierre 91
Stewart, Ian 65
subring 15
Subring Criterion 16
symmetric difference 12

Index

Tall, David 65
trisection of an angle 67
trivial ring 10
two-squares theorem 56

unique factorization
 domain (UFD) 37
unit 23

Vandermonde's
 determinant 84

Waring 33
Well-Ordering
 Principle 40
Wilson, Sir John 33
Wilson's Theorem 32

Yellow Brick Road 62

zero element 9